SEMINARS IN
CHIROPRACTIC

A QUARTERLY
SERIES

THERMOGRAPHY

VOLUME 1, NUMBER 1, WINTER 1990

SEMINARS IN
CHIROPRACTIC
A QUARTERLY
SERIES

THERMOGRAPHY

James Christiansen, PhD

Professor and Chairman, Department of Physiology, The National College of Chiropractic, Lombard, Illinois

Geoffrey Gerow, DC, DABCO

Private Practice of Chiropractic, Buffalo, New York

Series Editors

Dana J. Lawrence, DC

Editor, *Journal of Manipulative and Physiological Therapeutics*
Director, Department of Editorial Review and Publication
Professor, Department of Biomechanics and Chiropractic Technique
The National College of Chiropractic
Lombard, Illinois

Stephen M. Foreman, DC, DABCO

Adjunct Assistant Professor, Department of Diagnosis Postgraduate Faculty
Los Angeles College of Chiropractic
Whittier, California

WILLIAMS & WILKINS

Baltimore • Hong Kong • London • Sydney

Editor: Jonathan W. Pine, Jr.
Associate Editor: Marjorie Kidd Keating
Project Editor: Jennifer Conway
Designer: Dan Pfisterer
Illustration Planner: Lorraine Wrzosek
Production Coordinator: Anne G. Seitz
Cover Design: Dan Pfisterer/Mike Kotarba

Copyright © 1990
Williams & Wilkins
428 East Preston Street
Baltimore, Maryland 21202, USA

ISSN: 1047-1227

Printed in the United States of America

                                              90 91 92 93 94
                              1 2 3 4 5   6   7   8   9 10

Series Editors' Foreword

SEMINARS IN CHIROPRACTIC has been designed to provide the chiropractic practitioner with clinically and academically relevant information. We will present a mix of material: modern concepts on the topic in question as well as information and procedures that can be directly applied in a busy chiropractic practice. In a sense, this resembles the type of material presented at weekend educational seminars; the best of these programs have always reflected the need to provide practical information.

Each volume (four issues) will cover topics relating to such areas and disciplines as thermography, clinical biomechanics, radiology, research, and clinical diagnosis and management. The most current information needs of the chiropractic profession will be served.

We intend to cover a wide variety of topics; at the same time we hope to develop a reader-driven program that will enable us to be both flexible and responsive. Your comments and suggestions will be gratefully appreciated.

Dana J. Lawrence, DC
Stephen M. Foreman, DC, DABCO

Preface

Thermography has recently undergone tremendous growth in its application to medical science. From a technology once largely a military creation used to track troop movements by use of heat emissions, thermography has slowly evolved into a procedure that can objectively document the presence of a wide variety of human ailments. Some of the ailments for which thermography has been used include thyroid conditions, parathyroid problems, breast cancer, nerve root dysfunction, lumbar disk disease, and disorders of the sympathetic nervous system.

The chiropractic profession has played an important role in the evolution of thermography. The interest of our profession in thermal imaging procedures is logical, given that our history is replete with the development of heat-sensing instruments. The early 1900s found chiropractors using such devices as the neurocalometer to note bilateral temperature differences in the human body. Other devices soon followed, notably the thermeter and the tempograph. When the thermographic procedures developed by the military became public knowledge in the early 1960s, the chiropractic profession was quick to recognize the value of the new technology.

This first issue on thermography provides the interested chiropractor with scientific and practical information. In Part I, contributors discuss the historical development of thermography, the two major forms of this technology (electronic and liquid crystal) and the role the chiropractic profession has played in the newly emergent science. Insurance and medicolegal issues relevant to thermal imaging procedures are also addressed, and a sample deposition provides useful guidelines on developing a proper courtroom presentation.

In Part II, contributors present 14 case studies, which demonstrate the role this "new" technology is beginning to play in the objective documentation of certain disease and pathological states.

We hope that you, the reader, will find this material interesting and of value in your practice.

Dana J. Lawrence, DC
Stephen M. Foreman, DC, DABCO

Acknowledgments

We wish to express our thanks to Dana Lawrence for his diligent shepherding of this project. We thank those dedicated themographers who have continually promoted the use of thermography in research and clinical practice; their excitement prompted this book. We especially thank those practicing doctors who gave much of their time and effort to establish the International Thermographic Society and thus provide chiropractic thermographers a voice in the ever-expanding field of imaging technology. We also thank those thermographers who contributed so generously to this volume; without their efforts it could never have been completed. Finally, we specifically acknowledge Dr. W. N. Dudley, whose early publications encouraged the rest of us to enter the field.

We give special thanks to our families for their support and tolerance during the long days and late night hours of writing and rewriting.

James Christiansen, PhD
Geoffrey Gerow, DC, DABCO

Contributors

James Christiansen, PhD

Professor and Chairman
Department of Physiology
The National College of Chiropractic
Lombard, Illinois

Thomas J. Clay, DC, DABCT

Private Practice of Chiropractic
Edwardsville, Illlinois

William Dudley, DC, DABCT

Private Practice of Chiropractic
Howell, Michigan

Harold W. Farris, DC, FCTS

Private Practice of Chiropractic
Fresno, California

Robert Gaik, DC

Resident in Orthopaedics
The National College of Chiropractic
Lombard, Illinois

Geoffrey Gerow, DC, DABCO

Private Practice of Chiropractic
Buffalo, New York

Kim R. Hoover, DC

Private Practice of Chiropractic
Stuart, Florida

Jonathan Mueller, DC

Private Practice of Chiropractic
Crystal Lake, Illinois

William Pipher, DC

Resident in Orthopaedics
The National College of Chiropractic
Lombard, Illinois

Michael Poierier, DC

Associate Professor
Department of Diagnosis
The National College of Chiropractic
Lombard, Illinois

Robert Shiel, PhD

Associate Professor
Department of Diagnosis
The National College of Chiropractic
Lombard, Illinois

Contents

Part I
Overview

Part **II**
Case Studies

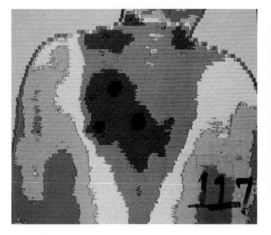

Figure 9.5

Figure 9.6

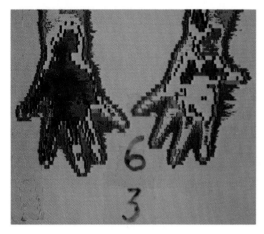

Figure 9.7

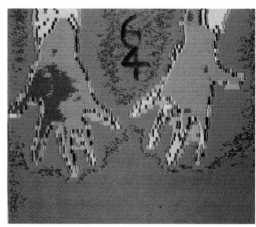

Figure 9.8

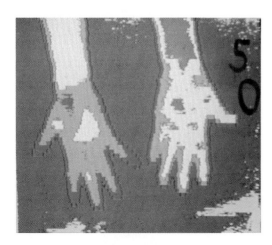

Figure 10.2

Figure 10.3

Part I

Overview

1

History of Thermography

James Christiansen, PhD
and Jonathan Mueller, DC

ANCIENT CONCEPTS OF BODY HEAT

From earliest recorded history, the ancient Greek philosophers and physicians (Plato, Aristotle, Hippocrates, and Galen) recognized and were fascinated by the relation between heat and life. The source of the body's heat was not questioned. Rather, the ancients speculated on the means whereby body heat was dissipated. Respiration was seen as an obvious cooling mechanism, primarily because expelled air felt warm (1).

Hippocrates noted temperature variations in different parts of the body. He considered increases in the "innate" heat of the body as the chief diagnostic sign of disease states: " . . . should one part of the body be hotter or colder than the rest, then disease is present in that part" (2). Indeed, Hipprocrates was said to have first felt radiant heat with the back of his hand and then confirmed it by smearing the area with wet mud and observing where it dried and caked first.

The ancient concepts of body heat were replaced by the discovery and development of the first air thermometer in 1592 by the astronomer Galileo. This crude instrument gave only gross indications of temperature changes, had no scale of measurement, and was influenced by atmospheric pressures. Sanctorius later modified the thermometer, devised his own, and described it in great detail (3). Boullian in 1659 modified Sanctorius's thermometer by introducing mercury into a glass tube. Later, Farhrenheit, Celsius, and Joule contributed the development of thermometric scales. The thermometric scale of Anders Celsius, known as the centigrade scale, gained general use in France and Germany, whereas the fahrenheit scale remained popular in England and the United States. Significantly, the thermometer was not routinely used to confirm or document internal body temperature, but passed into and out of prominence for the next 200 years.

Although fevers were much discussed during the 18th century, physicians did not routinely measure patients' temperatures, even though the mean healthy temperature had already been established by Becquerel and Brechet at 98.6°F (3). The first to establish and publish observations of body temperature and its variation in fever was Anton De Haden in 1754. Later James Currie also recorded temperature changes in febrile diseases. Wunderlich in the early part of the 19th century expanded on Currie's ideas. Over the course of his temperature studies, Wunderlich observed some 25,000 patients. Once his work was confirmed by others, the thermometer became by the late 1800s the standard oral measure of body temperature (4).

Claude Bernard regarded the nervous system as the regulator of all functions concerned with the maintainance of internal homeostasis. In the case of body heat, he believed that control was exerted through nerves, which not only caused vasoconstriction or vasodilation but also were responsible for corresponding local decreases or increases in metabolism (1).

At about the same time as Bernard's work, experiments on the effects of sectioning the spinal cord made at various levels were being conducted by Brodie, Chossat, and Broca (1). These experiments drew attention to the role of the sympathetic nerves in heat regulation. Broca's demonstration that a frontal lobe lesion in the brain produced hypothermia stimulated the search for brain centers that controlled physiological functions, including temperature regulation. By the early 1930s, considerable anatomical and physiological evidence pointed to the hypothalamus as the site of body temperature regulation.

Today we know that areas in the hypothalamus are indeed involved in temperature regulation. Generally, control of heat loss is centered in the anterior or preoptic hypothalamus, and control of heat conservation mechanisms lies in the posterior hypothalamus. These areas receive neural input from extrahypothalamic and peripheral thermoreceptors located throughout the body.

Not all early medical thermometry was limited to the measurement of internal body temperature. Recognizing variations of temperature in different parts of the body and believing that those variations were reflected on the skin surface, Spurgin in 1857 constructed a "thermoscope" (5). By comparing surface temperatures, he was able to diagnose breast tumors, having discerned that the heat of the tumor was several degrees higher than surrounding tissue. Satisfied with the results obtained with the instrument, he recommended the thermoscope be used in diagnosing and treating tumors and diseased joints. The use of skin surface temperatures for diagnosis was also advocated by Sequin who published a text on the use of surface temperature, but the concept did not gain acceptance and would not be pursued until a century later (5).

CHIROPRACTIC THERMOLOGY

In an effort to use surface temperature differentials to aid diagnosis, D.D. Palmer used the sensitive dorsal surface of the hand to locate "hot boxes" along the spinal column. These areas of increased heat relative to surrounding tissue were thought to identify inflamed nerves caused by impingement or subluxation (6). This technique, although subjective and unreliable owing to the variable sensitivity of the diagnosing physician, has been taught to chiropractic students since the birth of the profession.

In the early 1920s Doss Evins, an electrical engineer and student of B.J. Palmer, developed a heat-sensitive device which could locate paraspinal areas of increased heat semiquantitatively (6). The first neurocalometer (NCM), patented in 1925, consisted of two probes containing bimetalic strips connected to a meter. The NCM was used to compare two points on either side of the spinal column. If the two contralateral points were the same temperature, the meter needle would remain centered, but if one probe passed over a hyperthermic area ("hot box") the needle would deflect.

Although B.J. Palmer was thoroughly convinced of the efficacy and importance of this new analytic tool, others were not so enthusiastic and disavowed its use in chiropractic analysis. Their anxieties stemmed not so much from the physical principles of the tool itself, but from the wide-ranging physiological and diagnostic implications, such as understanding heat regulations in the body, derived from its application. Some enthusiastic supporters claimed the neurocalometer could identify the existence, location, and extent of a vertebral subluxation. Much of the intraprofessional rancor that exists within chiropractic today can be traced directly to the introduction and advocacy of NCMs by B.J. Palmer and his disciples (6).

Even though early NCMs were cumbersome and inaccurate because of technological inadequacies, the idea of quantifying surface temperature abnormalities or asymmetries remained appealing. Whether in support of this basic need for chiropractic documentation or to capitalize on a growing profession in search of technical expertise, many companies began manufacturing and merchandising devices designed to assess body surface temperature or cutaneuos blood flow. Produced in direct competition with NCMs, most devices were based on similar design and engineering concepts, whereas others invoked unique physical and physiological principles.

The simplest differential temperature devices incorporated two thermocouples and a meter in a simple hand-held probe. Two such devices, the Nervoscope and the Thermeter, are examples. Permanent recording of the meter deflection was achieved by coupling the instrument to a strip chart recorder. Thus, line deflections reflected "hot boxes." One company developed a motorized device to move the thermocouples at a constant rate down the spine, thus providing better correlation between thermal record and anatomical segment. Another company developed the ChiroProbe, which separated the two thermocouple probes, thus allowing differential temperature assessment at sites distal to the spinal column. Several of these instruments are still available today.

Other attempts to perfect surface temperature recordings resulted in the DermaThermograph (7), the Synchrotherme (8), and more recently the Visitherm (9) and the TyTRON (R. Titone, personal communication, 1988). These instruments detect skin surface temperature but record only single points or lines of temperature. Albeit accurate, the information is too limited to provide the complete thermal profile necessary for adequate thermographic diagnosis.

Concurrent with these developments in paraspinal surface temperature measurement was the birth of the Visual Nerve Tracer (VNT), designed by Adelman to perform the same spinal analysis based on different physiological principles (10). The basis of the VNT was that local paraspinal hyper- or hypothemia caused or was the result of increased cutaneous blood flow in the region. The VNT was a photoelectric reflection meter; its two paraspinal probes emitted visible light, filtered to pass only wavelengths around 556nm. This wavelength is strongly absorbed by the hemoglobin present in red blood cells. Reflected light was then detected by a photocell in each probe, and the contralateral probes were compared electronically to produce a line graph showing deflections at areas of altered cutaneous blood flow. The advantage of the noncontact VNT was that it did not touch or irritate the skin and therefore could not produce artifactual results. Its design was theoretically sound in that it detected blood flow changes rather than a secondary thermal signal. However, technical problems in its use and calibration prevented the VNT from being accepted by clinicians. Adelman also was one of the first researchers to use infrared imaging in diagnosis. He used near infrared film to photograph what he termed hyperthermias due to subluxations.

Criticism over the use of thermal detectors and the way in which Chiropractors interpreted findings, led many critics to label these instruments "a menace to the patient's health" (11). Yet many proponents believed thermal detectors were a shortcut to diagnosis and placed great reliance on them. Their unreliability and lack of scientific documentation has prevented their widespread acceptance in Chiropractic practice. Nevertheless, their use is still advocated, and case studies using these single trace "thermometers" continue to appear in the literature.

General use of surface temperature analysis fell into disuse until the advent of modern medical thermography in the early 1960s. Reference to chiropractic use of thermography does not appear until the early 1970s (12–15). Jenness (16) recognized the importance of thermography to Chiropractic as an adjunct in determining structural disrelationships and neuropathic processes. With these exceptions the disappearance of thermal mapping from chiropractic literature is reminiscent of the gaps in medical temperature research over the past centuries.

THERMAL IMAGING

The study of infrared radiation (IR) first began with optical experiments by della Porta at the end of the 16th century (17). Two centuries later Sir William Herschel, using spectroscopy, discovered that the sun emitted infrared rays. This discovery and its relation to light did not become clear until the mid-19th century, when Herschel's son, Sir John Herschel, a pioneer in the field of photography, produced on paper the first "thermograph" (17).

At about the same time, Langley developed the bolometer, a device capable of detecting radiant heat from living objects at a distance of up to 400 meters (18). The instrument and its potential uses were not further explored or developed, however, until a century later. Although different approaches to producing visible IR pictures had been attempted in the mid-1800s by Becquerel, Golay, and Czerny (19), each technique had produced thermograms that lacked sufficient temperature discrimination for their intended uses.

With the advent of World War II, IR technology advanced but was restricted to military use only. Czerny's evapograph was improved by adding newly developed thermistors which were coupled to an image-detecting device. The result was a new instrument which could detect, although crudely, troop movement, ground terrain, and ship movements at night. Several years later a Canadian physician, Dr. Ray Lawson, requested access to this military instrument for possible experimental medical applications.

In 1957 Lawson observed that the presence of breast cancer was reflected by increased skin temperature. His initial investigations were aided by R. B. Barnes and the development of the Barnes thermograph. The device consisted of a thermistor bolometer which detected the heat emitted and transformed it into electrical signals. The signals then illuminated a gas discharge tube which glowed with an intensity proportional to the radiation detected by the thermistor. The light was then reflected onto photographic film to produce the thermogram.

More advanced equipment soon followed, which greatly decreased the 10–15 minute screening time required by the origional Barnes thermograph. One such instrument, the Pyroscan, produced a thermogram in approximately 30 seconds—still considered too long by most medical practitioners (20).

In the late 1960s the Swedish firm AGA produced the AGA Thermovision, whose ability to generate a TV-like image on a cathode ray tube was a technological breakthrough. It permitted instantaneous observation and simultaneous recording of thermal patterns and thermodynamic processes in the human body (21). Its advanced IR technology and increased efficiency of operation rapidly made it the state of the art in medical thermography. Several other companies quickly produced instruments of comparable quality, and modern thermography grew rapidly.

Until the late 1970s, little documented evidence existed to show that thermographic images had a direct relation to clinical or diagnostic findings other than for breast disease. Lack of proper training and understanding of the equipment and protocol led to improper use of the technology and misinterpretations of the thermograms. These drawbacks resulted in thermography eventually being discredited by much of the medical profession.

Despite this poor prognosis for its medical use, sophisticated electronic IR equipment was introduced in the early 1970s for industrial purposes. One important advancement was the development of a colored isotherm modality. The isotherm delineates the heat picture as a pattern of color-coded isothermic bands capable of distinguishing variation in temperature gradients as small as 0.1°C. (21). This was a major breakthrough for thermography as a highly effective screening modality using body skin surface temperatures. Several researchers began using the technology to evaluate back and spinal temperatures (18), finding that patients with spinal injuries frequently displayed thermal asymmetries paraspinally. Their research led to a resurgence of the use of thermography for other than breast disease. Contemporary infrared thermography records a total thermal picture of the back or other anatomical areas without physical contact or irritation of any kind. Comparisons, paraspinally or distally, can be made not only for individual points or single lines, but also for entire regions of the body's surface (12). Simple electronic manipulations or sophisticated computer enhancement can isolate and analyze specific areas of interest for diagnosis.

At the same time IR detectors were being developed in the early 1960s, another technique of differential temperature was being explored, which employed cholesteric liquid crystals to produce colored thermograms of surfaces. In Liquid Crystal Thermography (LCT), once these materials are applied to the surface, they change color, reflecting the temperature (22). The surface can then be photographed directly, with the color patterns indicating specific temperatures.

Liquid crystal thermography did not become practical for medical diagnosis until 1967 with the technical improvement of encapsulation (23). In this technique, a self-contained thermal detector, which can be placed directly on the skin, changes color in response to any variations in skin surface temperature. Like electronic thermography, the entire thermal pattern of a specific region is displayed and interpretation is not dependent on single-point comparisons. The technical simplicity and cost of LCT have gained it wide acceptance among practitioners who use it for the majority of medical thermograms performed today.

CONCLUSION

The phenomenon of heat radiating from the body's surface has been recognized since the time of Hippocrates. D.D. Palmer in the late 1800s used this ancient concept to detect "hot boxes" along the spinal column for chiropractic diagnosis. In an effort to document subluxations and their postulated surface temperature abnormalities, various instruments were developed to detect, display, and quantify skin surface temperatures. Because of technological problems, lack of research, and poor understanding, the wide-ranging physiological research and diagnostic implications of these early instruments were never fully accepted by the chiropractic profession.

Modern developments in differential surface temperature measurement and analysis parallel early attempts in chiropractic analysis. With the advent of medical infrared thermography in the early 1960s, many technical problems were overcome. Figure 1.1 depicts the history of clinical thermal detection within the medical and chiropractic professions.

Even though thermography has now become an accepted medical imaging modality and the literature is replete with case studies and reviews (3,24–27), few scientific studies with statistically analyzed findings on groups of subjects have been published. Such research is surely the next phase of thermographic development.

Medical and chiropractic thermography for neuromusculoskeletal diagnosis and analysis is still primarily used to determine the paraspinal (and peripheral) thermal asymmetries associated with spinal pathology, long ago recognized by DD Palmer. Our understanding of the physiological processes responsible for Palmer's "hot boxes" has evolved significantly, as have our

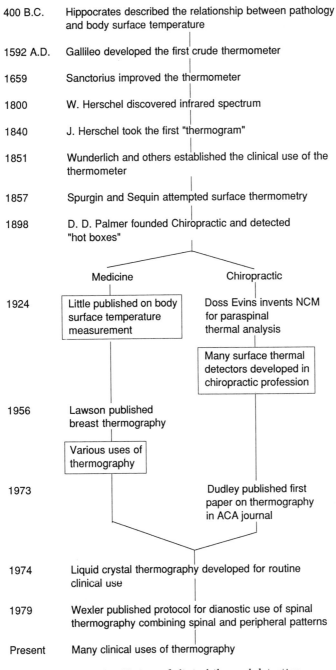

Figure 1.1. History of clinical thermal detection.

techniques for analyzing them. Thermal changes far removed from the spine are now recognized as frequently being caused by spinal trauma. The chiropractic profession today has at its disposal a technology that enables it to objectively document some of the physiological processes associated with subluxations and altered nerve functioning, processes that are fundamental to chiropractic healing philosophy.

References

1. Lomax E: Historical development of concepts of thermoregulation. In (ed): *Body Temperative—Modern Pharmacology—Toxicology.* New York, Marcel Dekker, 1979, vol 6.
2. Adams F: *The Genuine Works of Hippocrates.* Baltimore: Williams & Wilkins, 1939.
3. Gershon-Cohen J: A short history of medical thermography. *Ann NY Acad Sci* 122:4–11, 1964.
4. Haggard HW: *The Doctor in History.* New Haven, Yale Univ. Press, 1934.
5. Haller JS: Medical thermography—a short history. *West J Med* 142:108–116, 1985.
6. Dye AA: *The Evolution of Chiropractic.* Richmond Hall Inc, 1939.
7. Kimmel EH: The derma thermograph. *J Clin Chiro* 1:78–86, 1969.
8. Haldeman S: First impressions of the synchro-therme as a skin temperature reading instrument. *J CCA* April, pp 7–8, 1970.
9. Stillwagon KL, Stillwagon G: *Visi-Therm 747.* Videotape presentation. Monongahela, PA, 1984.
10. Novick ND: The VNT photo-electric instrument. *J Clin Chiro* 2:78–83, 1969.
11. Kimmel EH: Electro analytical instrumentation. *ACA J Chiro* 6:S33–44, 1966.
12. Dudley WN: Thermography and the body. *ACA J Chiro* 7:S30–32, 1973.
13. Dudley WN: Thermography: a clinical study. *ACA J Chiro* 8:S30–31, 1974.
14. Dudley WN: Facial thermography and adjustment. *ACA J Chiro* Aug, pp 54–56, 1974.
15. Dudley WN: Extremity thermography and low back pain. *ACA J Chiro* 11:S29–30, 1977.
16. Jenness ME: The role of thermography and postural measurement in structural diagnosis. *NINCDS Mono No 15* DHEW Pub. No. (NIH) 76–998. Washington, DC, DHEW, 1975.
17. Putley EH: The development of thermal imaging systems. In Ring EFJ, Phillips B (eds): *Recent Advances in Medical Thermology.* New York, Plenum Press, 1982.
18. Rask MR: Thermography and the human spine. *Orth Rev* 8:73–82, 1979.
19. Gershon-Cohen J: Medical thermography. *Sci Am* 216:94–102, 1967.
20. Curcio BM, Haberman J: Infrared thermography: a review of current medical application, instrumentation and techniques., 1971.
21. Ryan J: Thermography. *Australas Radiol* 13:23–26, 1969.
22. Koopman DE: Cholesteric plate thermography: the state of the art. *Ann NY Acad Sci* 181:475–480, 1980.
23. Hobbins WB: Differential diagnosis of pain using thermography. In Ring EFJ, Phillips B (eds): *Recent Advances in Medical Thermology.* New York, Plenum Press, 1984.
24. Lawson RN: Thermography: a new tool in the investigation of breast lesions. *Can Ser Med J* 13:517, 1957.
25. LeRoy LP, Bruner WM: Effects of electrical stimulation on the thermographic pattern in the human patient with chronic pain syndrome. In Gautherie M, Albert E (eds): *Biomedical Thermology* New York, Alan R. Liss Inc., 1982.
26. Christiansen J: Thermographic physiology. In Rein H (ed): *The Primer on Thermography.* Sarasota, FL, H Rein, 1983.
27. Clark RP: Human skin temperature and its relevance in physiology and clinical assessment. In Ring EFJ, Phillips B (eds): *Recent Advances in Medical Thermology.* New York, Plenum Press, 1984.

2

Modern Medical Thermography

James Christiansen, PhD

Following publication of the first medical thermogram by Lawson in 1956, the use of thermography in diagnosis grew quickly. Military infrared technology was declassified in 1956, and several researchers immediately began using the new devices for determination of skin surface temperature. Gershon-Cohn et al. (1) in the U.S. and Lloyd-Williams (2) in Britain both initiated studies. The first symposium on thermography was sponsored by the New York Academy of Sciences (NYAS) in 1963. Although papers were presented on several potential medical uses of thermography, the emphasis remained on breast cancer detection. The conclusion of the symposium was that research into the effective diagnostic uses of thermography should continue (3).

Several extensive prospective studies followed. Dodd et al. (4), Hobbins (5), Isard (6), Amalric (7), and Gautherie et al. (8) were pioneers in this research and found significance in its diagnostic abilities. Others found the techniques to be less satisfying, however. Some (9) found too many false-positive thermograms suggesting breast cancer, and others (10,11) found false-negatives, resulting in missed cancer diagnosis. In an extensive nationwide study initiated in 1973 and sponsored by the American Cancer Society and the National Cancer Institute, thermography was compared to mammography and biopsy for diagnostic ability. Results were so poor and inconclusive that the Breast Cancer Diagnosis Demonstration Project (BCDDP) dropped thermography from its evaluation protocol in 1977 (12). Other studies, both prospective and retrospective, have appeared over the years, but thermography's reputation as a cancer diagnosis modality had been severely jeopardized. Whether owing to inadequate training, unrecognized technological sensitivity, or incomplete understanding of thermophysiologic regulatory mechanisms, thermography has not regained the stature it once held in diagnostic imaging.

At the same time that the early thermographic ventures were occurring in the United States, European researchers were using thermography extensively for breast cancer diagnosis and other studies. Use of thermography grew slowly but more steadily in Europe, perhaps because claims were less vocal or the economic reimbursement system was different, but not because expectations were lower. Several researchers and institutions in France collaborated on large-scale breast cancer studies (11). Gautherie has used these early breast cancer studies in performing extensive research on mathematical models and computer design (12) as well as clinical research (13).

Following the 1964 NYAS report (3), the European thermologic community held a symposium in 1966 and organized the European Thermographic Society, which first met in 1971 (14). This organization sponsored the first thermographic journal, *Acta Thermographica.* Japanese thermographers held their first meeting in 1967. European and Japanese thermologists continue to explore innovative ways to use this technology in medical diagnosis and research.

In spite of the virtual dismissal of the use of breast thermography, many American clinicians continue to believe in its diagnostic efficacy, and symposia devoted exclusively to breast thermography remain well attended (15). In fact, some studies have suggested that many previously diagnosed false-positive breast thermograms may have been subclinical, and that thermography may have been overlooked as perhaps the most important early screen modality (15).

AMERICAN THERMOGRAPHIC SOCIETIES

The American Thermographic Society was founded in 1971 and began holding annual meetings to report progress in the field of medical thermology. The proceedings of their meeting at Johns Hopkins University in 1975 was edited by S. Uematsu and published as *Medical Thermography, Theory and Clinical Applications* (16). This small book was considered the "bible" of thermography for many years and remains the best introduction to its basic principles and uses.

The American Thermographic Society changed its name to the American Academy of Thermology in 1985 to emphasize a general approach to all thermobiologic interests, and in 1986 it introduced *Thermology* as its official journal. Adopted by the European Thermographic Society in 1987, the journal today represents general medical thermography in most western countries.

In an effort to distance themselves from the controversy between breast cancer and thermography, independent thermographers specializing in other types of diagnosis organized the Academy of Neuromuscular Thermography in 1985. This group focuses on the use of thermography in other than breast disease, and it certifies medical thermographers for expertise in these protocols.

Chiropractic thermographers organized as the International Thermographic Society (ITS) in 1983. From its inception, ITS sponsored educational symposia and funded thermographic research at several institutions. This international, interprofessional organization established the American Board of Clinical Thermology (ABCT) in 1985, the first independent examining board to test candidates for certification in chiropractic thermography. The responsibilities of this board were transferred in 1988 to the American Chiropractic Board of Thermography (ACBT), under the auspices of the newly formed American Chiropractic College of Thermology, which was based within the American Chiropractic Association's Council on Diagnostic Imaging. The ITS is now organizing a new board for general recognition and certification of thermographers and technicians.

The California Thermography Society also represents a significant number of chiropractic thermographers. The group recently changed its name to the International Academy of Clinical Thermology to broaden its appeal and increase its recognition. It examines thermographers and awards certification to qualified physicians and technicians.

DEVELOPMENT OF THERMOGRAPHY

Thermography has been used to document and aid in the diagnosis of many conditions other than breast disease. Earliest interest appears to have centered on vascular problems, on which research continues today (17–20). Cerebrovascular syndromes also have been documented with thermography (21–24). These conditions frequently provide clear thermographic evidence, but often they are inconclusive because the bilateral pathology on which many are based generally cannot be documented using standard techniques for thermal asymmetry.

Some researchers have suggested that thermography can be used to measure blood flow (25) and to determine basal metabolic rates in the evaluation of dietary requirements (26). Others have used thermography in conjunction with acupuncture to demonstrate a physiological change associated with the needling process (27,28).

Thyroid and parathyroid diseases also have been suggested for thermographic diagnosis (29–31), on the premise that hyperactivity of either of these organs coincides with increased blood flow to provide metabolites. Artificial cooling with alcohol spray produces hypothermia; rapid blood flow will quickly return the temperature to precooling levels, whereas hypoactivity will be accompanied by a slower return to normal. Although certain researchers are convinced of the technique (WM Dudley, personal communication), the procedure has not been widely used.

Despite early studies involving a variety of conditions, thermography is used most to diagnose lower back pain with radiculopathy and myofascial trauma. Opinions on why this occurs vary, but some of it may be explained by the high incidence and economic impact of both syndromes (32), as well as the extent to which both are difficult to objectively diagnose and document.

SCIENTIFIC STUDIES

Although thermography is widely used in diagnosis and many case studies and clinical reports have been written, relatively few scientific studies exist in the literature that quantitatively and statistically validate the use of thermography in clinical diagnosis.

Neuromuscular thermographic diagnosis is based on the premise of symmetry of cutaneous blood flow in the right and left sides of the body. This assumption was first demonstrated thermographically in surgically proven abnormal lumbar disc patients. Researchers found that 80% of proven disc patients had a positive thermogram. The thermographic interpretation was subjective, however, and did not indicate the degree of asymmetry used to define a positive thermogram. Similarly, Tshauer (33) found back-pain patients to be asymmetric and normal subjects to be symmetric, but his clinical criterion was simply "pain," and he specified no thermal criteria.

Most early studies of musculoskeletal thermography evaluated symptomatic subjects using thermography alone or in combination with other techniques (34–36); others, however, evaluated asymptomatic subjects for thermal asymmetries. Silverstein et al. (37) showed that thermal asymmetries of the back did not exceed 1°C in normal males; Kamajian and Tilley (38) also stated that asymmetries of 1° were unusual. Others have more recently revised the value downward to 0.63°, and Uematsu (39,40) has compiled tables of thermal asymmetries for all body surfaces, showing typical values of less than 0.5° asymmetry for any body surface.

Uematsu (16) was the first to establish a protocol for defining an abnormal thermogram. He described a 1.0° asymmetry in 25% of the area viewed as his criterion for thermal abnormality. Since then the same criteria have been expanded to include a 1° asymmetry in 25% of the cutaneous dermatome or area of nerve distribution. Uematsu's definitions addressed the asymmetry of views or body parts, whereas the current definitions incorporate an aspect of diagnosis; if there is an apparent thermal asymmetry, then the diagnosis is supported. Misuse of this current standard often leads to conflicting impressions. An irritation of the fifth lumbar nerve root, for example, may display as a thermal asymmetry of the lateral thigh, with no additional findings. Uematsu's criteria would show this to be an abnormal "thermogram", whereas current use of the criteria might lead to an interpretation of 'positive' for L5 nerve root irritation because the majority of the *view* is asymmetric; it might also, however, be interpreted as insufficient, since less than 25% of the *entire* dermatome is positive. This ambiguity in interpretive standards is one of the drawbacks for the acceptance of diagnostic thermography. Furthermore, few "blind" studies have investigated the interobserver reliability based on subjective interpretations of thermographers (41,42). This limitation is perhaps the most glaring omission in current thermographic research.

Another of the difficulties in establishing criteria for positive or abnormal thermograms arises from the original studies which compared normal subjects to those having surgically confirmed disc lesions. The more recent studies, which have determined the variability of skin temperature in normal subjects, will permit a restructuring of the definitions to more accurately reflect an abnormal thermogram versus a positive diagnosis. Mild nerve irritations may cause neurovascular changes in certain portions of the nerve distribution that may not be detected by other diagnostic techniques. The result would be a highly sensitive procedure that could detect mild as well as severe neural stimulation. Such detection could contribute to diagnosis but would not be interpreted as a positive diagnosis for any particular condition. A normal, symmetric thermogram, suggesting no neurological insult or vascular response, would be highly suggestive of a negative diagnosis, since almost any neural stimulus will result in some vascular response (see Chapter 5).

The evaluation of a new diagnostic modality is usually performed against some reference standard. Thermography images a unique parameter of diagnostic criteria, namely, the physiological response to neural stimulation of both the sensory and sympathetic nerves. This places thermography in the position of being evaluated to as either a clinical impression or an anatomic imaging procedures, neither of which is an appropriate reference. It has been compared to orthopaedic evaluations (19,43), computed tomography (CT), myelography, electromyography (EMG) (44), magnetic resonance imaging (MRI) (45) and somatosensory evoked potentials.

In a comprehensive review of many of the available studies assessing the accuracy, validity, and predictive value of thermography compared to other diagnostic tests, Meeker and Gahlinger (46) found thermography to be highly sensitive, with a high negative predictive value, but less specific and with a lower positive predictive value. It compared favorably to myelography, EMG, and CT, with accuracies of 87%, 74%, and 83%, respectively. Sensitivity exceeded 90% in twelve of the fourteen studies analyzed. The importance of this high correlation of thermography with other diagnostic tests is heightened by the fact that they measure different phenomena. Myelography, CT, and MRI primarily image anatomical structures, whereas EMG records changes in motor nerve activity. Alterations in these parameters may or may not produce functional disorders. Thermography, on the other hand, "images" physiological changes resulting from pathophysiological stimuli to the sensory or sympathetic nerves, thus providing complementary information to most other diagnostic tests.

CHIROPRACTIC REPORTS

Despite its rapid acceptance among many persons in the Chiropractic profession, thermography only recently was endorsed by the profession at large through its inclusion in the ACA Council on Diagnostic Imaging.

More than a decade elapsed between first reports of thermography in the Chiropractic literature and a smattering of case studies and reports in trade magazines (47–50). More recently, several reports have been published in chiropractic journals (46,51,52) and a new publication devoted to thermography, written primarily by Chiropractors, has released its first issue.

Although few studies exist on the use of thermography in chiropractic practice, several chiropractors have written educational manuals for physicians and technicians (53–55). Thermography has a definite place in the diagnosis and monitoring of chiropractic patients. The history of the profession has always included surface temperature monitoring (Chapter 1), and recent advancements in this modern technology will soon make thermography a standard laboratory test requested by practicing Chiropractic physicians.

References

1. Gershon-Cohen J, Berger SM, Haberman JD, Barnes RB: Thermography of the breast. *Am J Roentgenol* 91:919–926, 1964.
2. Lloyd-Williams K: Thermography in the prognosis of breast cancer. *Bibl Radiol* 5:62–67, 1969.
3. Dodd GD, Wallace JD, Freundlich IM, Zermeno A: Thermography and cancer of the breast. *Cancer* 23:797, 1969.
4. Hobbins WN: Thermography, highest risk marker in breast cancer. *Proc Gynecol Soc for the Study of Breast Disease* pp 267–282, 1977.
5. Isard, HJ. Thermographic 'edge sign' in breast carcinoma. *Cancer* 30:957, 1972.
6. Amalric R, Brandone H, Robert F: Infrared thermography of 2,000 breast cancers. *Acta Therm* 3:46, 1978.
7. Gautherie M, Gross C: Contribution of infrared thermography to early diagnosis, pretheraputic prognosis and post-irradiation follow-up of breast carcinomas. *Medica Mundi* 21:135, 1976.
8. Threatt, B, Norbeck, JM, Ullman, NS, Kummer R, Roselle PF: Thermography and breast cancer: an analysis of a blind reading. *Ann NY Acad Sci* 335:501, 1980.
9. Moskovitz M, Milbrath J, Gartside P: Lack of efficacy of thermography as a screening tool for minimal and stage I breast cancer. *N Engl J Med* 295:244, 1976.
10. Moskovitz M, Fox SH, del Re Brun R, Milbrath JR, Bassett LW, Gold RH, Shaffer KA: The potential value of liquid-crystal thermography in detecting significant mastopathy. *Radiology* 140:659, 1981.
11. Baker LH: Breast cancer detection demonstration project: five year summary report. *CA* 32:194, 1982.
12. Gautherie M: Temperature and blood flow patterns in breast cancer during natural evolution and following radiotherapy. In Gautherie M, Albert E (eds): *Biomedical Thermology.* New York, Alan R. Liss Inc, pp 21–64, 1982.
13. Gautherie M: improves system for objective evaluation of breast thermograms. In Gautherie M, Albert E (eds): *Biomedical Thermology.* New York, Alan R. Liss Inc, pp 897–905, 1982.
14. Aarts NJM: Presidential address. *Bibl Radiol* 6:ix–xiv, 1975.
15. Gautherie M, Haehnel P, Walter J-P, Keith L: Improved survival rate of breast cancer patients with an early diagnosis based on initial thermal assessment and subsequent follow-up. In Gautherie M, Albert E, Keith L (eds): *Thermal Assessment of Breast Health.* Boston, MTP Press, pp 180–201, 1983.
16. Uematsu S (ed): *Medical Thermography: Theory and Clinical Applications.* Los Angeles, Brentwood Publishers, 1976.
17. Spence VA, Walker WF: An assessment of thermography in arterial disease. In Ring EFJ, Phillips B (eds): *Recent Advances in Biomedical Thermology.* New York, Plenum Press, pp 337–344, 1984.
18. Miki Y, Kawatsu T, Matsuda K: Thermographic venography in inflammatory lower leg nodules. In Gautherie M, Albert E. (eds): *Biomedical Thermology.* New York, Alan R. Liss Inc, pp 439–443, 1982.
19. Pochaczevsky R, Pillari G, Feldman F: Liquid crystal contact thermography of deep venous thrombosis. *AJR* 138:717–723, 1982.
20. Stess RM: Thermographic evaluation of pedal diabetes. *Second Opinion* 1:15, 1988.
21. Capistrant TD: Thermographic facial patterns in carotid occlusive disease. *Radiology* 100:85–89, 1971.
22. Karpman HL, Kalb IM, Sheppard JJ: The use of thermography in a health care system for stroke. *Geriatrics* *:96–105, 1972.

23. Wood EH, Friedman AP: Thermography in cluster headache. *Res Clin Stud Headache* 4:107–111, 1976.
24. Kudrow L: A distinctive facial thermographic pattern in cluster headache—the 'chai' sign. *Headache* 25:33–36, 1985.
25. Miyake H, Fujimasa I, Iwatani M, Atsumi K: Development of a thermographic skin blood flowmetry system. In Ring EFJ, Phillips B (eds): *Recent Advances in Biomedical Thermology.* New York, Plenum Press, pp 227–234, 1984.
26. Van J: Body heat—thermography has appeal, but value? *Chicago Tribune* June 8, Sec 6, p 1, 1986.
27. Liao SJ, Liao MK: Acupuncture and tele-electronic infra-red thermography. *Acupunct Electro-ther Res* 10:41–66, 1985.
28. Ernst M, Lee MHW: Sympathetic vasomotor changes induced by manual and electrical acupuncture of the Hoku point visualized by thermography. *Pain* 21:25–33, 1985.
29. Samuels BI: Thermography: a valuable tool in the detection of thyroid disease. *Radiology* 102:53–62, 1972.
30. Samuels BI: The present status of parathyroid thermography. *JAMA* 25:907–908, 1975.
31. Samuels BI, Dowdy AH, Lecky JW: Parathryoid thermography. *Radiology* 104:575–578, 1972.
32. Anderson EF, Hegstrum JHA, Carboneau GJ: Multidisciplinary management of patients with chronic low back pain. *Clin J Pain* 1:85–90, 1985.
33. Tshauer, ER: The objective corroboration of back pain through thermography. *J Occup Med* 19:727–731, 1977.
34. Karpman HL, Knebel A, Semel CJ, Cooper J: Clinical studies in thermography II. Application of thermography in evaluating musculoligamentous injuries of the spine—a preliminary report. *Arch Environ Health* 20:442–447, 1970.
35. Raskin JM, Martinez Lopez M, Sheldon JJ. Lumbar thermography in discogenic disease. *Radiology* 119:149–152, 1976.
36. Rask MR: Thermography of the human spine; study of 150 cases with back pain and sciatica. *Ortho Rev* 8:73–82, 1979.
37. Silverstein EB, Bahr GJM, Katan B: Thermographically measured normal skin temperature asymmetry in the human male. *Cancer* 36:1506–1510, 1975.
38. Kamajian GK, Tilley P: Thermography of the back of asymptomatic subjects. *J Am Osteopath Assoc* 74:429–431, 1975.
39. Uematsu S: Symmetry of skin temperature comparing one side of the body to the other. *Thermology* 1:4–7, 1985.
40. Uematsu S: Thermography of cutaneous sensory segments in patients with peripheral nerve injury: skin temperature stability between sides of the body. *J Neurosurg* 62:716–720, 1985.
41. Uricchio JV, Walbroel CE: Blinded reading of electronic thermography. In (ed): *Academy of Neuro-Muscular Thermography: Clinical Proceedings.* Encino, CA, Academy of Neuro Muscular Thermography, pp 47–53, 1986.
42. Sherman RA, Barja RH, Bruno GM: Thermographic correlates of chronic pain: analysis of 125 patients incorporating evaluation by a blind pane. *Arch Phys Med Rehabil* 68:273–279, 1987.
43. Wexler CE: Lumbar, thoracic and cervical thermography. *J Neur Orthop Surg* 1:37–41, 1979.
44. Weinstein SA, Weinstein G: A clinical comparison of cervical thermography with EMG, CT scanning, myelography and surgical procedures in 500 patients. In (ed): *Academy of Neuro-Muscular Thermography: Clinical Proceedings.* Encino, CA, Academy of Neuro-Muscular Thermography, pp 44–46, 1985.
45. Goldberg G: Thermography and magnetic resonance imaging correlated in 31 cases. In (ed): *Academy of Neuro-Muscular Thermography: Clinical Proceedings.* Encino, CA, Academy of Neuro-Muscular Thermography, pp 54–58, 1985.
46. Meeker WC, Gahlinger PM: Neuromusculoskeletal thermography: a valuable diagnostic tool? *J Manipulative Physiol Ther* 9:257–266, 1986.
47. Pireno AA, Ricciardi DO: The accuracy and objectivity of computerized thermographic examinations. *Am Chiro* June, p14, 1986.

48. Arbiloff B: The current state of high resolution thermography in the United States and Europe. *Dig Chiro Econ* Sept/Oct, p22, 1986.
49. Cannon L: The validation of thermography. *Am Chiro* Feb, p12, 1987.
50. Kneebone WJ, Grand L: A correlation of lateral flexion lumbar spinal radiographs and thermograms on chiropractic patients: a pilot study. *Dig Chiro Econ* May/Jun, p76, 1988.
51. Diakow PRP: Thermographic imaging of myofascial trigger points. *J Manipulative Physiol Ther* 11:114–117, 1988.
52. Gerow G, Christiansen J: Thermographic imaging of rat feet following complete sciatic section. *J Manipulative Physiol Ther* (in press) 1989.
53. Chapman GE: *Neuromuscular Thermography,* Chula Vista, CA, CTA Publishers, 1984.
54. Chapman GE, Britt BA: *Thermographic Diagnostic Manual.* Chula Vista, CA, CTA Publishers, 1984.
55. Silverman HL, Silverman LA: *Clinical Thermographic Techniques.* Clayton, GA, Rabun Chiropractic Clinic, 1987.

3

ELECTRONIC THERMOGRAPHY

James Christiansen, PhD,
and Geoffrey Gerow, DC, DABCO

ELECTROMAGNETIC SPECTRUM

Electronic thermography is a technique by which radiant energy emitted from a surface is detected and imaged within the visible range. This radiant energy is greatest in the infrared range of the electromagnetic spectrum.

Figure 3.1 depicts the entire electromagnetic spectrum, including wavelengths in the gamma, ultraviolet, visible, infrared, and radiofrequency ranges. Shorter wavelengths indicate greater energy of the wave (1).

All solid and liquid materials absorb as well as reflect radiant energy, depending on the molecular composition and surface characteristics of the material. Visible "light" in the range of 400 nm–700 nm (0.4–0.7 μm) wavelength is perceived by the human eye and converted to "color" by the central nervous system. Although the wavelength perceived is usually reflected from the surface, whereas the other visible wavelengths are absorbed by the object being seen, this need not be the case, such as when a metal object is heated sufficiently to emit "red" or even "white" light. In these instances, the vibrational motion of the molecules within the material releases sufficient heat to produce radiant energy. The temperature of the material is another measure of the amount of vibrational energy.

Some materials characteristically reflect the radiant energy that strikes them; others absorb this energy and convert it to another form (such as heat) or reemit the energy at a different wavelength. The fate of incoming radiant energy is partly dependent on the wavelength of the energy itself. As explained above, whether a given wavelength in the 400nm–700 nm range is absorbed or reflected depends on the "color" of the object. This same concept applies to wavelengths in other portions of the electromagnetic spectrum as well. It is

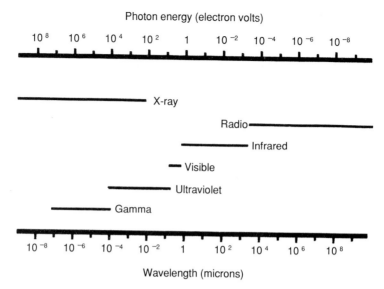

Figure 3.1. Electromagnetic spectrum showing the wavelength and energy of the various portions of the spectrum.

the basis of infrared (IR) photography, in which reflected IR rays are captured on film and converted to chemical energy to "develop" a picture. Specific IR wavelengths are reflected by some materials, others are absorbed. Infrared film can absorb and convert wavelengths in the near visible range, but it does not absorb longer wavelengths. (If it did, it would constantly be "exposed" by the IR radiation emitted by the camera itself!) The reflected radiation is captured by IR film, not the wavelengths emitted by the object being photographed. Infrared thermography, on the other hand, captures the energy spontaneously emitted from an object, rather than reflected rays.

Vibrational molecular motion ceases at absolute zero (0°Kelvin; −273°C). Any temperature above 0°K produces molecular motion, and the amount of radiant energy released is proportional to the fourth power of the absolute temperature, as expressed by the Stefan-Boltzman equation:

$$E = \epsilon \sigma T^4$$

E = total emissive power (Watts per square meter; W/M²)
ϵ = emissivity, a characteristic of the material; $\epsilon_{max} = 1$
σ = a proportionality constant = 5.672×10^{-8} WM^{-2} °K^{-4}
T = absolute temperature in degrees K.

1.

Another law of physics, Wein's law, recognizes that as an object is heated, its spectrum of emitted wavelengths also changes, owing to expansion and molecular alterations of the material itself. The law suggests that an optimal wavelength (λ_{max}) of emission exists for any given temperature, and that peak wavelength changes as the temperature changes. The mathematical equality is expressed as:

$$\lambda_1 T_1 = \lambda_2 T_2 = \text{constant.}$$

2.

Max Planck combined these concepts into a single equation which represented all wavelengths:

$$E_\lambda = C_1 \lambda^{-5} \, (\text{Exp} \, (C_2/\lambda T) - 1)^{-1}$$
$$C_1 = 3.732 \times 10^{-10} \, \text{Watts-cm}^2$$
$$C_2 = 1.4387 \, (\text{cm-}°\text{K})$$
$$\lambda = \text{wavelength}$$
$$T = °\text{K}$$

3.

The most important equation for practical thermography is an expanded form of the Stefan-Boltzman formula:

$$\text{Energy transfer (W/M}^2) = \epsilon\sigma \, (Tb^4 - Te^4)$$
$$T_b = \text{body or object temperature}$$
$$T_e = \text{environmental temperature}$$

4.

Equation 4 emphasizes that energy transfer from the body to the environment is dependent on the temperature difference. If there is a significant thermal difference between the body and the environment, there will be a great deal of energy transfer. Physiologically this heat loss will be detected and homeostatic mechanisms initiated which may influence the thermogram. Near equilibrium conditions are necessary to permit the determination of "normal" energy transfer to the environment.

Figure 3.2 shows that all objects emit energy in the $1-25$ μm (IR) spectrum.

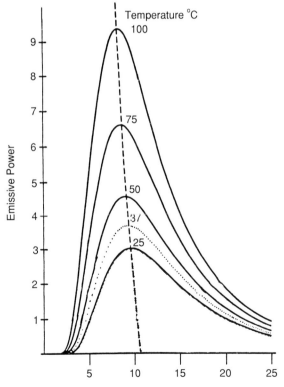

Figure 3.2. Wavelength versus emissive power. The wavelength of maximum emissivity decreases with increasing temperature.

The total emitted energy is the area under each curve. At increasing temperatures a greater proportion of radiant energy is emitted at or near the λ_{max}, and the wavelength of maximum emitted energy shifts slightly toward shorter wavelengths. As the temperature increases even more, beyond that depicted in the figure, the spectrum of energy emission will approach the visible range and the object will "glow" red with emitted energy. For thermographic purposes, however, it is sufficient to note that an object at 37°C (human body temperature) emits energy over a wide spectrum, with a maximum energy output at 9.3 μm. The curves differ at every point, and the area under the curves, representing total radiant energy output, is proportional, whether measured from 1–5 μm, 1–10 μm or 1–20 μm.

BLACK BODIES

Again, objects may absorb or reflect radiant energy that strikes them. If an object reflects all radiant energy in the visible spectrum, its color is white; if it absorbs all visible radiation, it is black. Similarly, if an object absorbs all IR radiation, it is called a "black body." A black body, receiving radiant energy, must convert that energy into another form or reemit it, sometimes at another wavelength. This is the principle behind a black object becoming warmer in direct sunlight. It absorbs energy in the visible spectrum and converts it to heat energy. It will then radiate more energy in the IR spectrum. If an object has high reflectance, such that little energy is absorbed, then it will have low emissivity; if an object has low reflectance, it will absorb radiant energy, ultimately reemit it, and thus have a high emissivity (ε, epsilon).

Human skin has a high emissivity in the IR range, nearly equal to that of a black body. In equations 1–4, emissivity (ε, epsilon) approaches unity and can generally be taken as 96–100%. Furthermore, since the emissivity factor is constant for most conditions of healthy skin, it has little significance in interpretation. Certain conditions exist, however, in which the emissivity of human skin may change significantly. In these cases, although the numerical value is indeterminate, the possible role of emissivity should be considered. A change in emissivity of as little as 0.02 will represent an error in temperature measurement of nearly 0.5°C.

RADIATION DETECTORS

As seen in Figures 3.1 and 3.2, at 37°C the most energy radiated from an object is in the 2–25 μm IR range. Infrared film can image only wavelengths up to 1.2 μm in the near IR range, and consequently is inappropriate for photographing radiant energy. Several detectors have been developed to quantify and record the radiant energy flux in the middle and far IR ranges. Figure 3.3 shows a series of radiation detector materials and their effective spectral ranges. As a practical matter, owing to cost, availability, sensitivity, and efficiency, only the alloys of indium-antimonide (InSb) and mercury-cadmium-telluride (HgCdTe) are typically used in medical thermography instruments. Although InSb has an upper wavelength cutoff at less than six μm, below the E_{max} for objects at 37°C, and representing only 2% of the total

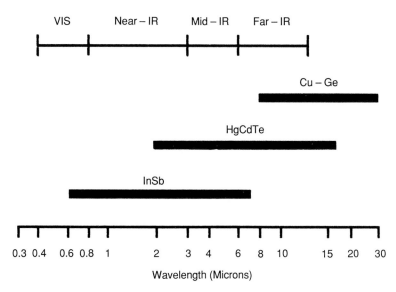

Figure 3.3. Standard infrared detectors and their ranges of sensitivity. Indium-antimonide (InSb) and mercury-cadmium-telluride (HgCdTe) are commonly used IR detectors, but copper-doped germanium (Cu-Ge) is rarely used.

energy emitted, its sensitivity and response time make it an adequate IR detector. HgCdTe can detect IR radiation at 9.3 μm, the E_{max} for 37°C, but its sensitivity and speed reduce its desirability. In any event, since the detector responds to the total radiant energy focussed on it, and the total energy emitted is proportional to the area under the curve in Figure 3.2, it is not critical to capture the entire spectrum. High sensitivity and response speed may outweigh detectivity for any given detector.

The function of an IR detector is to convert the incident radiation into another form of energy, which is proportional to the IR energy and quantifiable in usable units. Modern detector materials convert IR energy into electrical energy which can be amplified and measured using conventional technologies. Detector materials respond to increasing temperature by increasing molecular vibrations. This molecular activity releases electrons in proportion to the radiation absorbed. Obviously, at environmental temperatures electrons are constantly being released, producing a constant electrical signal. Excess radiant energy, above background, will increase the signal only in proportion to the thermal increase. Thus, for small temperature differences, the signal to noise ratio will approach unity. To maximize the efficiency of a radiation detector, the spontaneous signal must be reduced, such that the incoming IR energy will produce the maximum possible signal. The optimal signal to noise ratio is usually achieved by cooling the detector to as low a temperature as possible. Many thermography units cool the detector with liquid nitrogen (LN₂, − 196°C). Others use the evaporation of compressed argon (Ar) gas. Still other modern units cool the detector with piezoelectric refrigeration devices.

It is impossible to collect radiant energy from an entire object in a single view, because the radiation detector converts incident energy into an electrical signal rather than a photographic image. Instead, discrete points are "imaged" in sequence and the entire object view is compiled as an accumulation of discrete "point sources" on a cathode ray tube, or more commonly a TV-type monitor. Individual "point sources" must have sufficient energy to excite the detector to release electrons. The lower limit of spot size sufficient to excite the detector is dependent on detector material, temperature, distance, angle, and so forth. The maximum spot size is also important from a practical view. Accumulation from too large an area will obscure the necessary thermal and spatial discrimination. Empirical evidence suggests that for medical thermography, a maximum spot size of 2 mm diameter is sufficient for thermal and spatial resolution. Modern thermography units typically subtend an angle of about 1.3 mrads. This means that they capture an instantaneous image from a spot size of approximately 1 mm diameter at a distance of 1 m from the detector. This spot size is well within the resolution limits required for practical thermographic imaging.

The amplified electrical signal from the IR detector can be recorded and quantified in a number of ways. Initially the electrical response to thermal input is displayed on the monitor as intensity. A strong signal, representing high radiant energy from the source, will produce a bright (white) point on the screen, against an unlighted (black) background. The accumulated points present a composite image. The output signal can be broken into stepwise proportions of the maximum detectable signal, and each step can be assigned a unique color using standard electronic manipulations. Thus, a multicolored image, representing discrete thermal ranges (isotherms), can be displayed on a color monitor. This display is frequently stored on videotape for later display and analysis.

Certain difficulties arise with the use of this technology. The relationship between intensity of display and temperature is indirect and not linear. The display of most black and white (gray) thermography units is calibrated in isotherm units. No electrical signal is seen as a black screen at zero isotherm, whereas one isotherm unit represents 100% intensity of electrical input and a bright white image. Appropriate calibration of the display screen will result in a continuum of shades of gray, from black (0 isotherm) to white (1.0 isotherm). Figure 3.4 demonstrates that with uniform sensitivity on a 10°C range, a 10% change in signal intensity (white) will represent a 1°C difference in temperature. The same 10% change in signal intensity using a 5° range will represent 0.5°C. Incorrect calibration of the display monitor will result in inaccurate interpretation of isothermal level and temperature differences. Figure 3.5 shows that if the Gray scale is too narrow, most of the intensity change will occur over a narrow isotherm range (0.4–0.6), and the temperature change will be overestimated. If the Gray scale is too broad, covering only 40% of the possible signal intensity, the temperature differences will be underestimated.

Similar technical difficulties accompany the use of colored isotherm bands. Most thermographers interpret thermal asymmetries based on the differences in color between contralateral body surface areas or adjoining

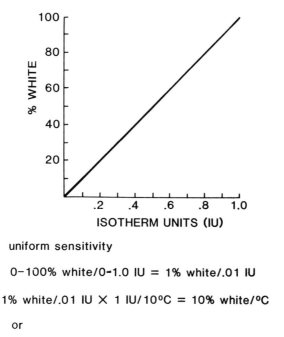

uniform sensitivity

0-100% white/0-1.0 IU = 1% white/.01 IU

1% white/.01 IU × 1 IU/10°C = 10% white/°C

or

1% white/.01 IU × 1 IU/5°C = 10% white/0.5°C

Figure 3.4. Electronic Gray scale at correct settings showing linear relationship between illumination and isotherm units and the conversion of isotherm units to temperature.

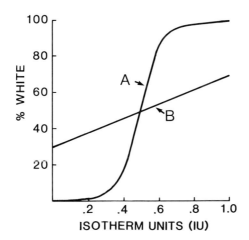

too much contrast
70%white/.2 IU × 1 IU/10°C = 35%/°C

too little contrast:
40%white/1.0 IU × 1 IU/10°C = 4%white/°C

Figure 3.5. Electronic Gray scale at incorrect settings, showing too much contrast (**A**) and too little contrast (**B**). Incorrect interpretation of temperature will result from either error.

regions. Since each colored band represents a range of temperature, that interpretation may be underestimated or overestimated. Figure 3.6, illustrates this point. Each color represents 0.1 isotherm or approximately 1°C at 10°C full scale. Points A and B, displaying different colors, may be as little as 0.1°C apart, whereas B and C, the same color, may differ by as much as 0.9°. Similarly, A and C, in two adjoining color bands, are approximately 1° different, whereas B and D, also in adjoining colors, are as much as 1.9° apart. This technical drawback easily can be overcome using modern, computer-linked equipment which can report accurate temperatures for individual points. Most clinicians in routine applications do not, however, quantitatively evaluate thermograms.

Each discrete spot imaged by the detector can be digitally stored in computer language. The technique permits storage, recall, and quantitative comparisons of any point or region within the stored image. It also allows quantitative comparisons between images if appropriate precautions, such as a constant thermal reference, are included. This quantitative approach will certainly be the important new direction in electronic thermography.

PRACTICAL CONSIDERATIONS

Apart from the actual capture of radiant energy from a specific solid angle or arc, several practical considerations influence the medical utilization of thermography. First, energy is emitted from a point source in all directions. The greatest amount of energy is emitted in a line normal (perpendicular) to the surface being imaged, but the amount of energy captured is, obviously, dependent on the distance of the detector from the surface. This observation is analogous to the capture of visible light by the eye. The closer we are to an object, within limits, the more light our eye will capture and the clearer the image perceived. For this reason it is most practical to be as near the object as

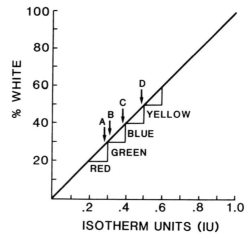

Figure 3.6. Electronic color scale showing the accumulation of signal intensity into discrete colors.

possible. Since the spot size has, however, been determined to be approximately 1 mm at a distance of 1 m, an approach closer than 1 m may jeopardize resolution. Distances greater than 1m will increase the object's surface area contributing to the thermal signal and also decrease spatial and thermal resolution. Moreover, if the image occupies less than half the screen monitor, image clarity (spatial resolution) is being sacrificed. It is appropriate to use as much of the monitor screen as possible to improve resolution. Lastly, IR radiation is not absorbed by dry air and therefore has an almost infinite range. It is, however, absorbed by moisture in the air (humidity) and reradiated. Although of little consequence for medical thermography, this will reduce both spatial and thermal resolution as distance increases.

A second practical consideration for medical thermography is the angle of view for thermal detection. As stated earlier, most IR radiation is normal to the surface of the object; yet, with the exception of particularly smooth, flat surfaces, texture will provide microsurfaces radiating in all directions from the plane of the object surface. Skin is extensively textured and radiates at all angles. This enables imaging of standard views (anterior, posterior, lateral) rather than in a continuous arc around the body. However, as incident angle increases, the proportion of radiant energy directed toward the detector decreases. This is particularly true when the angle exceeds 40° (S or P). This consideration is important when temperatures on the periphery of the object are being imaged and evaluated. If the detector is not perpendicular to the plane of the object, the difference in incident angle for the two edges is the sum of both angles and may represent a significant distortion of thermal resolution.

Electronic, infrared thermography is not unique to the medical imaging profession. It is used in industry, research, theater, and weather forecasting, as well as by governmental agencies. Its use in diagnostic imaging, however, requires special training and understanding of both the equipment and the physiology involved in the production of the thermal image.

PRODUCT INFORMATION

A number of companies manufacture telethermographic equipment. Not all companies are listed below, yet the absence of any one company should not be considered a lack of trust in its product.

FLIR Corporation

The FLIR Corporation has been manufacturing thermographic systems for the military for years. Its interest in medical applications of thermographic equipment is more recent. The FLIR imager is a two-detector unit, and, as such, offers a distinct advantage over most thermographic cameras which have only one detector.

The FLIR model 110A radiometric thermal imager produces 140 IR lines per scan and 250 IR pixels per line. The image contains 35,000 data points and is digitized into high resolution. Stored in real time, the entire image is updated 30 times a second. It may be stored to a computer or frozen for evaluation. FLIR uses the addition of two black-body heat sources placed in front of the camera unit to obtain maximum specificity and calibration of the image evaluated. Up to 256 colors or 256 Gray scale levels are available.

Image analysis includes split screen images, spot temperature readout, vertical temperature profile, horizontal temperature profile, histograms of areas of interest, image averaging, color palette selections, normal and inverted video, isotherm, temperature in Celsius, Fahrenheit, or Kelvin, and electronic zoom. The system utilizes computer key board or mouse control. An optional 35mm camera attachment is available for the Polaroid freeze-frame unit. The detector consists of a two-element HgCdTe, patented serial scan, sensitive to the wavelength of 8–12 μm. The camera uses compressed argon to cool the detector and has a running time of 4 hours with the standard unit, and up to 24 hours with an optional setup.

The field of view is 26.7°H x 15°V. The rage of temperatures assessable is 10°–70°C.

Model 110A, with its two-detector method, provides the best overall resolution of a medical image we have seen to date. The product's medical software is, however, disappointing. Examination time is slowed considerably because the program does not accommodate multiple picture-taking and storage after exposure—important criteria in medical thermography. If FLIR were to overcome this barrier, its product could be the best overall system on the market.

Mikron Corporation

The Mikron Corporation has three models of thermographic equipment: 6T62, 6T63, and 6T66, only one of which—the 6T66—is marketed to the medical and biological community. Although the first two models could be used for medical thermography, their application is solely industrial. Model 6T66 can display up to 64 colors or grays; a temperature bar on the side of the image calibrates the colors. The machine is sensitive to a temperature of − 10° to 50°C. The unit has good resolution and, according to the company, is sensitive to .03°C.

Mikron offers the following features: autothermal focus; outline feature (will outline all areas at or above a preset temperature); color isothermal display (in this mode, only a certain temperature will be highlighted); optical zooming (provides zooming from 1–5 times the image size); average temperature display (allows person to map up to four separate zones for analysis and read the average temperature for each zone on the side); half run (a split screen on which one side shows a frozen image and the other a real time image); local point temperature display (up to 10 points may be taken for point temperature evaluation on a given image); and a subtraction process, which displays the difference between two images by comparison of the pattern with standard data.

Dorex Corporation

The Dorex Corporation produces two systems: the DCATS 386, using 32 Bits and 386 16/20MHz CPU, and the DCATS 286 16 BIT/286 10/12MHz CPU. Both machines are a single detector system composed of HgCdTe, sensitive to the 8–12 μm wavelength. Each unit can image up to 62,000 data points. The image is captured in real time but requires 4 seconds per scan. The unit is calibrated by an internal black body source, and the detector is cooled by liquid nitrogen. Sensitive to .1°C, the unit can display up to 256 hues, but on any individual image, 16 is the preferred number. The monitor can display up to 20 images simultaneously. An area for patient data is available with each image, which can be boxed and evaluated for symmetry, right to left. Space is provided for the patient's name, identification number, time, date, the distance of the camera from the object in question, the height and width in pixels of the boxed areas of interest, mean temperatures for the boxes in question, and the temperature range of the image viewed. The image has good resolution of 512 x 480 pixels, and the monitor is a flat screen to limit distortion. A limiting factor is the need to store each image at the time it is taken, since images are not given sequential identification numbers. Scans done in this manner would tend to increase the time needed to image the patient.

Agema Corporation

The Agema Corporation, formerly AGA, has developed a camera that uses a single detector made of InSb. The detector is sensitive to 3–5.6 μm of the spectral band, which is exceptionally sensitive even though it reads only 2.4% of the energy emitted from the human body. The camera uses a germanium lens with two rotating prisms; the first prism rotates at 180 rpm, the 2nd at 18,000 rpm. These prisms are uniquely synchronized to allow the infrared stimulus to strike the detector. The medical unit is sensitive from 28°–39°C. Although previous models had required liquid nitrogen as a coolant, the latest thermovision 870 uses a thermoelectric cooling device. The unit contains two built in temperature references. The gray tone image can be converted to color through the use of Discon, a scan converter that makes thermovision compatible with TV monitor and video cassette recorders.

The computer-aided thermography software has the following advantages: it can spot absolute thermal measurements up to five points in real time; use two isotherms for making selected areas; perform profile temperature measurements in the vertical, horizontal, and slanted positions; conduct a zone analysis of the absolute temperature of up to 7 boxes; subtract two images simultaneously on the screen; and magnify images.

The software program supplied with the system is especially designed to guide the technician through the thermographic evaluation process based on information obtained from the Academy of Neuromuscular Thermography. The program has prompts for which pictures are to occur next. The thermographic evaluation proceeds with the touch of a single button. With each patient exam, the software provides the opportunity to log patient identification, history, and other relevant data and stores this information with the exam. Images may be produced in a video color printer for hard copy; 35 mm slides and ink jet printing are options.

The National College of Chiropractic uses an upgraded AGA 720 unit, with Viewscan's Viewsoft and View Medical software programs (Ontario, Canada). We use both programs to advantage. The Viewsoft program allows up to 16 pictures to be taken with single button pressure and holds these pictures on screen with progressive numerical identification. Once the pictures have been loaded, they may easily be stored to the hard drive. When asked, the computer will reload the series as a unit for analysis by reading the first number of the series. Pictures may be calibrated to the absolute temperature probe and uniform assignment of colors and temperature demarcations made. The program allows for analysis of the data by zone analysis of stipulated areas of interest, allowing the clinician to obtain maximum temperatures, minimum temperatures, and mean temperature.

Hughes Corporation

Hughes Corporation has produced a number of thermographic devices, the most recent of which is the Model 7300 Probeye Thermal Video System. This solid-state infrared imaging system has an HgCdTe detector sensitive to 8–12 μm. The detector is cooled by a solid-state Peltier electric cooler. The unit has a focus range of 11 inches to infinity, with 240 infrared lines, 128 color levels, and 128 gray levels. Model 7300 captures images in real time with a rate of 30 frames per second. Although the machine's technology is impressive, we feel its medical software support needs improvement.

Inframetrics

Inframetrics Company of Bedford, Massachusetts, markets a basic thermography unit, the Model 520M, which it promotes as a "forensic special." This unit uses a liquid nitrogen-cooled HgCdTe detector and is sensitive to radiation in the 8–12μm region. The scanner does not have a lens, so its focal distance can go from 0 to infinity. It displays in black and white or digitized color, with 10 selectable thermal ranges, and has an isotherm function. Additional features include day, date, and time displays, patient information, and color code. Output from the basic unit can be displayed on a TV-type monitor, or it can be fed to a computer for data manipulation using available software. Inframetrics also has sophisticated equipment hooked directly to computers for data storage and retrieval.

Reference

1. Poole DO. Physical basis. In Raskin JM, Viamonte R (eds): *Clinical Thermography.* 1977.

4

Liquid Crystal Thermography

*Geoffrey Gerow, DC, DABCO,
and James Christiansen, PhD*

Liquid crystals are anisotropic organic molecules that have one or more mesophases between the crystalline and liquid states (1). First described in 1889 by two researchers (2,3), liquid crystals were of little practical value because they were difficult to handle. Their use in medical research originally was accomplished by applying a black base coat to the subject being thermographed, followed by a coating of liquid crystals. The subject was then photographed, and the liquid crystal material was washed away and discarded. Liquid crystal thermography (LCT) was not generally used in medical diagnosis until 1965, when the material was applied to a durable plastic sheet which could be reused on patients without difficulty. An innovation two years later in which the crystals were encapsulated into flexible plastic sheets virtually eliminated any problems in their handling (4).

Liquid crystals in their mesomorphic state are neither solid nor liquid; they are a cloudy, viscous material in which groups of individual molecules interact to form an ordered, structured group. Based on a system proposed by Friedel, liquid crystals may be arranged in three forms: smectic, nematic, and cholesteric (5). Medical thermography utilizes cholesteric liquid crystals, because they are a type of molecule that produces optically active crystaline structures (6). Cholesteric liquid crystals are named for the cholesterol or cholesterol derivatives contained in many of the optically active molecules.

The cholesteric liquid crystal molecules assume a parallel relationship, one to another, and stack together in a helical structure. The height of the stack that forms one rotation of the helix is the pitch (7). The helix may display either right-handed or left-handed pitch and demonstrate optical properties termed circular dichroism. Circular dichroism means that when white light strikes the crystalline structure, most wavelengths are transmitted and pass through the material, whereas certain wavelengths cannot pass through and are reflected. If the reflected wavelength is in the visible spectrum, the surface

31

takes on the color of that wavelength. The reflected wavelength is characteristic of the pitch of the crystal.

An interesting phenomenon occurs with the introduction of heat to liquid crystalline substances. As temperature rises, the pitch decreases owing to improved fit of the molecules as they increase vibrational energy. Since the reflected wavelength is directly proportional to the pitch, the resultant "color" of the material changes as temperature changes. Mathematically, the molar optical rotation of a liquid crystal material is $(a)^{YT} = 100aK$.

Where (a) is equal to the specific rotation, a is the measured optical rotation, T is the absolute temperature, and Y is the wavelength of polarized light. When the pitch is equal to the wavelength of a specific color, the liquid crystal will reflect this color (6). As the temperature rises, the pitch becomes smaller, and the colors reflected are those of blue-violet. As the temperature falls, the pitch increases, and the reflected color will pass through yellow and green, toward the red end of the visible range. Cholesteric liquid crystals are sensitive to a wide variety of temperatures between 20°C and 200°C, depending on the mixture of cholesteric materials. Mixing of cholesteric crystals provides sensitivity in the range of temperatures useful in medical imaging.

Modern LCT is performed using a thin elastomeric sheet approximately 0.25mm thick. The liquid crystal material encapsulates the sheet, which is attached to a frame which can be stretched over the contours of the body (8). Most units are mounted on an apparatus which allows the elastomeric sheet to be inflated and stretched to improve body contact. These thermographic units have a thermal sensitivity of approximately 0.2°C, such that a change in temperature in excess of 0.2° will produce a change in the reflected wavelength and color (9), as well as spatial resolution of 3mm (4). A color reference chart is attached to each thermography unit indicating the actual temperature associated with each color being reflected. The image produced by contact of the thermographic unit with the subject is then photographed for a permanent record.

Certain vital technical considerations govern the use of LCT in medical diagnoses. First, the relationship between temperature and reflected color is not linear. This is partly because of the specific cholesteric compounds used to produce the material and partly because the transition in color depends on the stacking of the cholesteric molecules, which does not occur in discrete steps; the transition is continuous. Interpretation of colors will depend on the manufacturer and even the specific screen being used. Figure 4.1 depicts the transition of colors in thermography units manufactured by Flexi-Therm, Inc. The shift in temperature required to change from, say, tan to red brown using a 30° plate (screen) is approximately 0.5°C (28.8–29.3), whereas the same transition using the 32° plate is nearly 1.5°C (30.0–31.4). Changes in color do not reflect the same degree of temperature shift for each screen used. Some manufacturers have described a 1°C calibration between discrete colors of their liquid crystal spectrum, even though the effect of temperature on the transition, as we have said, is continuous. Any comparison of the thermal image with the reference provided can only be subjectively interpreted.

Temporal resolution of cholesteric liquid crystals also is not absolute; the heat capacity of the elastomeric sheet and the cholesteric material may influence the final temperature and color of the thermogram. Moreover, a

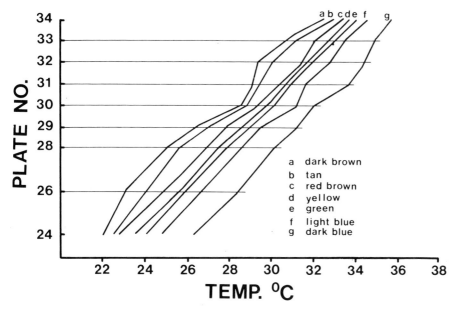

Figure 4.1. Flexi-Therm color transitions.

hysteresis or difference occurs in the packing of the liquid crystals, depending on whether the final, steady-state temperature is arrived at by warming or cooling the material (4). A warm thermography screen applied to a subject will equilibrate at a different reflected wavelength than a cool screen. This means the thermography unit must always be cooled to below the temperature of the object being thermographed before performing thermography.

Another problem arises from the pressure sensitivity of the material. Pressure will influence the stacking of the cholesteric molecules and may produce a different pitch from that caused by temperature alone. The influence of pressure is, however, much less than that of temperature and is minimal for most medical applications.

Liquid crystalline material is susceptible to ultraviolet light and requires appropriate care. Most cholesteric molecules are degraded by continued exposure to ultraviolet light, but the degradation is gradual and does not become evident immediately. Consequently, LCT units have a limited lifetime, which can be extended by storage in a light-tight container.

The advantages of LCT are that it is portable and convenient to use. Yet this simplicity often leads to abuse, because important protocols necessary for accurate thermography are omitted. Liquid crystal thermograms provide a direct reading of the actual temperature of the area viewed, since the color reflects the actual temperature. Electronic thermography provides discrete bands of color representing temperature bands or ranges, but the continuous transitions of liquid crystal colors permit a more accurate, immediate interpretation. Use of more than one liquid crystal screen permits thermal resolution to very small absolute temperatures.

Liquid crystal technology has existed for nearly as long as infrared thermography. Its accuracy and reproducibility have been well documented in both medical and industrial applications. It has been used to diagnose many disorders, including temporomandibular joint disease (10), carpel tunnel syndrome (11), subluxations (12), low back pain (13), spinal root compression syndromes (14), reflex sympathetic dystrophy, and upper limb entrapment syndromes (15). If there is any concern over LCT, it is that its availability has tended to increase its misuse by persons not well trained. Its continuing value as an important diagnostic modality should minimize this concern, however, as technical proficiency assumes an increasingly vital role in diagnostic accuracy.

PRODUCT INFORMATION

Liquid crystal products have certain advantages over electronic thermography systems. They are considerably lower in price and more portable, and they have absolute temperature values inherent in the colors represented within the image. In some electronic thermography units, the latter feature is an add-on.

We believe the chief disadvantage of LCT is the increased time and effort required to obtain thermographic studies and the inability to perform post-screening analytic computations. Three manufacturers of LCT products are described below.

Flexi-Therm, Inc.

Flexi-Therm, Inc. produces the Mark IIB. The system includes eight thermo-flex detectors, which range from 24°–33°C, a photographic recording system, storage for the detectors, a carrying case, a foot pump inflator, a patient identification system, and protective latex covers. Training is also provided. The Flexi-therm unit features a high brightness of colors, a .6°–1.0°C interval between colors, and six colors for each of the 8 detectors. Each 11 x 14 inch detector has a 5°C range and a window. The Mark IIB, which weighs 7 lbs, records with a Polaroid or 35 mm camera.

QMAX/BIOTHERM

The QMAX/BIOTHERM company produces the Q-scan system. Weighing 11.5 lbs, the system has a moderate brightness of colors, a 1°C interval between colors, four detectors, and six colors per detector. Each detector has a 7° range, two thermal windows, and measures 9.5 x 14 inches. Q-scan has a fan and cord, and records with a Polaroid. A stand with support arm is available through Biotherm, and the company can connect interested persons with training services.

MEDTECH/NOVATHERM

Novatherm, manufactured by MEDTECH/NOVATHERM, has a moderate brightness of colors, .8°C per color interval, and eight detectors, each of which has six colors, a range of 4°C, two windows, and measures 12.5 x 16.5 inches.

Each detector, with frame, weighs 6–7 lbs and records with a Polaroid camera. The company provides training.

Several characteristics of the three products are worth noting. As we explained earlier, changes in color do not reflect the same degree or temperature shift for each screen used. In the case of Q-scan and Novatherm, the two windows featured on each detector will cause some colors to represent two different temperature ranges. The convenience of not having to change detectors as often tends to be offset by this occurrence, which may be confusing to the technician or interpreter. Biotherm offers a stand to support its thermal imaging device. This may be of benefit to the technician who need not hold this comparatively heav'r unit for all the positions.

In general, a unit with a smaller thermal range for six colors will provide better resolution than spreading six colors over a wider thermal range. Flexi-Therm's Mark IIB offers the tightest sensitivity to temperature differences (e.g., .6°–1.0°C) per color of the three machines. One drawback, however, is that the Mark IIB has a nonlinear relationship between the color and gradations of temperature difference. This lack of standardization of colors representing a specific thermal difference may be somewhat confusing to the operator.

References

1. Huang C: Liquid crystals, numerical models of mesophases. *Nature* 332:781, 1988.
2. Lehmann O. Fliessende Kristalle. *Z Phys Chem* 4:462, 1889.
3. Reintzer F. Zur Kenntnis des Cholesterins. *Monatsh Chem* 9:421–425, 1889.
4. Fiesch U: Techniques of liquid crystal thermography in medicine. In Engel JM, Flesch U, Stuttgen G (eds): *Thermological Methods*. Fed Rep Ger, VCH Publishers, 1985, pp 45–62.
5. Fergason J: Liquid crystals. *Sci Am* 211:77–85, 1964.
6. Goodby J: Optical activity and ferroelectricity in liquid crystals. *Science* 231:350–355, 1986.
7. Tonegutti M, Acciarri L, Racanelli A: Cholesteric liquid crystals. Supplement 3 to *Acta Thermogr*, Fundamentals of Contact Thermography in Female Breast Diseases pp 11–15.
8. Meyers P: Vacuum and inflatable contoured liquid crystal Flexi-Therm contact thermography. *Radiol Today* 1:307–309, 1981.
9. Katz S, Masterfano T: A preliminary report on the clinical testing of a new elastomeric liquid crystal film. *Ann NY Acad Sci* 335:484–488, 1980.
10. Finney J, Holt C, Pearce K: Thermographic diagnosis of temporomandibular joint disease and associated neuromuscular disorders. *Academy of Neuromuscular Thermography Clinical Proceedings*. New York: McGraw Hill, 1985, pp 93–95.
11. Herrick R, Herrick S, Puohit R, Smith L: Liquid crystal thermography in the detection of carpal tunnel syndrome. *Academy of Neuromuscular Thermography Clinical Proceedings*. New York, McGraw Hill, 1985, pp 78–82.
12. Reynolds C: Liquid crystal thermography, its use to detect subluxation. *Today's Chiropractic* June pp 18–19, June 1985.
13. Rubal B, Traycoff R, Wing K: Liquid crystal thermography, a new tool for evaluating low back pain. *Phys Ther* 62:1593–1596, 1982.
14. Pochaczevsky R, Feldman F: Contact thermography of spinal root compression syndromes. *Am J Neuroradiol* 3:243–250, 1982.
15. Nakano K: Liquid crystal contact thermography in the evaluation of patient with upper limb entrapment neuropathies. *Neurol Orthopaed J Med Surg* 5:97–102, 1984.

5

Thermographic Anatomy and Physiology

James Christiansen, PhD

HEAT PRODUCTION

Heat is continuously produced in the body as a by-product of metabolism. Three major factors determine the rate of heat production: basal metabolic rate, specific organ activity, and local muscular activity.

A person's basal metabolic rate is largely a function of predetermined hypothalamic set points in the central nervous system (CNS) and hormonally controlled metabolism, influenced primarily by secretions from the hypothalamus and thyroid glands. Metabolic heat production remains constant throughout the thermoneutral range of environmental temperatures, principally through reflex changes in the body's insulative and heat distribution mechanisms. These reflex vascular changes may alter the thermal profile as seen thermographically, but they have little importance in the interpretation of diagnostic thermography. Instead, diagnostic thermography is based for the most part on thermal asymmetries, right to left, and the relation of discrete surface areas to their surroundings. Although thermography has been used in an attempt to measure basal metabolic rate (1), it has little relevance for clinical diagnosis and will not be considered here.

The metabolic activity of individual organs may have a profound effect on the overall temperature of the body. For example, metabolic heat production following a meal has been termed "food inducible thermogenesis" (2), and is related to increased metabolism during digestion. Disease processes within visceral organs may also stimulate metabolism. Bacterial and intrinsic pyrogens may produce fever, such that body temperature may increase dramatically. The heat generated by visceral metabolism is transferred directly to the blood passing through these organs. The blood is mixed, on passage through the lungs and heart, and redistributed throughout the body. It has been well documented that none of the heat produced locally in an organ is conducted through the overlying somatic tissues to be reflected on the body's surface (3,4). The heat produced by visceral activity is evenly distributed to the surface of the body for dissipation and cannot be used for thermographic diagnosis. This is not to say that thermography is of no value in diagnosis of visceral pathology. Many visceral pathologies produce pain sensations at unique sites in the somatic tissues. These "referred" pain areas have been shown to respond thermographically, secondarily reflecting the visceral pathology (5,6). The mechanism of this pain and thermal referral will be discussed later in this chapter.

Muscular activity has an obvious effect on heat production. Exercise may increase oxygen consumption tenfold, with a corresponding increase in heat production. The increased heat production of excess muscle activity may be visualized thermographically, but unless the increase is due to a disease process, it has little relevance for diagnostic thermography. The protocol for proper diagnostic thermography specifies that the patient must be at equilibrium or steady state with his or her environment for proper thermal imaging. Muscle exercise is not consistent with the necessary equilibrium conditions.

In certain instances increased muscular activity may have implications for diagnostic thermography. The thermal profile surrounding a myofascial trigger point reflects, to some degree, the metabolic activity of the involved muscle (7). The increased metabolism of a muscle in continuous contraction will produce heat which is carried away by venous drainage. This vascular drainage may carry the warmed blood vertically to near the surface of the body, where some of the heat may be lost to the environment. The cutaneous hyperthermia, is however, more likely caused by restricted blood flow through the muscle in spasm. Thus, core temperature blood is redirected into collateral vessels in the skin, producing a warm image over the site of a trigger point (7,8).

HEAT LOSS

Changes in metabolic activity or heat production are not the basis of clinical, diagnostic thermography. Rather, thermography reflects the distribution of heat throughout the body and regulation of that heat loss, especially at the surface, in the skin. Metabolic heat, whether from visceral or somatic structures, is brought to the body core through the large veins and vena cava, where it is mixed and distributed through the aorta and arteries. Peripheral and CNS receptors detect minor changes in temperature and alter the peripheral blood flow to maintain a constant core temperature by adjusting the vascular networks to dissipate excess heat from the surface of the body. Core temperature blood is directed through superficial vascular networks, which are controlled to permit the loss of appropriate amounts of heat to maintain core temperature. It is the distribution of blood in these superficial vascular nets that is imaged by thermography.

The actual mechanisms of heat loss from the body are conduction, convection, evaporation, and radiation. Heat flows from one solid (or fluid) to another through the process of conduction. Heat loss from the body by this method is minimal under normal conditions, accounting for less than five percent of metabolic heat production. Our body surface makes little contact with other solids, allowing minimal surface area for conduction. Even clothing makes little total contact with the body, and most clothing materials are poor conductors.

Heat transfer by convection is the movement of thermal energy in a fluid. As a mechanism of heat loss from the body, it is represented by the warming of the microenvironment of the skin (conduction), followed by the replacement of that warm air through currents. Less than 15 percent of metabolic heat is actually lost from the body using this mechanism, especially if the body is clothed, since the warm air is trapped within the insulative layers. Convection of heat is, however, an important heat transfer mechanism in the overall process of thermographic diagnosis. Heat transfer from one part of the body to another (e.g., core to periphery) is accomplished through fluid movement (blood flow). Blood, warmed by metabolism in both visceral and somatic structures, is convected through the vascular tree and transferred, first to the body core and then to areas of lower temperature. Thus, convection is the major mechanism of heat transfer *within* the body (9).

Evaporation is the conversion of water to water vapor, which requires heat energy. This process occurs on all body surfaces, but it is an important heat loss mechanism in the lungs and respiratory tract; significant amounts of water are vaporized and exhaled. Evaporation has little importance in thermographic imaging when the subject is at equilibrium with his or her environment. If the subject is not at steady state, however, and increased heat loss becomes a physiological requirement, evaporation of perspiration may become a significant heat loss mechanism and produce thermographic artifacts.

The major mechanism of heat loss from the human body is radiation, which accounts for some 60% of the total loss. Heat energy is converted to electromagnetic, radiant energy, which is emitted from the body in the infrared (IR) range. Core temperature blood is brought to the cutaneous vascular networks, where the heat energy is converted to radiant energy and transmitted to the environment. At body temperature this energy radiation is maximal at around 9nm wavelength, which can penetrate less than 1mm of tissue without being absorbed and reconverted to heat. Furthermore, the vascular networks which carry the heat to the surface lie only 1–2mm beneath the surface (9). Consequently, heat loss by radiation occurs only from the surface of the body and cannot reflect any deep-lying heat sources. It is this IR radiation that is detected in electronic thermography.

ANATOMICAL CONSIDERATIONS

Medical thermography uses the heat loss from the body to assess physiological homeostasis. Heat loss is a regulated physiological function which is controlled through vasomotor activity in cutaneous vascular nets. Thermography does not detect pathology, rather it images the vasomotor response that may occur secondarily to pathological processes. This thermographic imaging of cutaneous blood flow is based on several anatomical and physiological assumptions.

First, thermography is based on the recognition that the human body is segmented. This segmentation, readily apparent in embryonic development, is lost in the adult, with certain exceptions. Embryonic myomeric segmentation is not recognizable in the skin, somatic musculature, or visceral organs of the adult; it does, however, remain evident in the segmental nature of the osseous spinal column and in the neuroanatomy of the spinal cord and spinal nerves. Individual spinal nerves of unique segmental origin can be traced to specific distribution areas within the body. Although these nerve distribution areas overlap to some degree, their anatomical and physiological areas of distribution have been well established (10,11). The somatic segmentation of the adult can, in fact, be extrapolated from the distribution of the segmental nerves from the spinal chord.

Sensory nerves originate as cell bodies within the dorsal or posterior root ganglion. They send axons centrally to the lateral areas of the dorsal horn of the spinal column and distally to visceral and somatic structures. Somatic sensory axons project to the skin and musculature through mixed function spinal nerves and terminate in free nerve endings in the cutaneous as well as muscle layers. The dendritic terminals may branch into sensory arbors, or they may be closely associated with blood vessels with synapse-like boutons (12,13).

These somatic sensory neurons may send collateral branches to sympathetic ganglia, where they appear to form synapses with postganglionic sympathetic neurons (14,15), or they may bifurcate, sending separate axon branches to discrete somatic structures (16,17). The cutaneous distribution of these sensory neurons from a single spinal segment form a unique map on the surface of the body, the dermatome. A similar, three-dimensional distribution may be presumed for the muscles (somatome), but it is less well characterized.

Most visceral sensory axons also have their cell bodies within the dorsal root ganglion. The axons pass through the sympathetic ganglionic chain by way of the white ramus and terminate as dendritic free nerve endings within the various visceral organs. Somatic sensory neurons transmit nociceptive afferent information to the CNS for processing, but they may also send axon branches to various sympathetic ganglia, where they form axodendritic synapses with postganglionic neurons emanating from these sympathetic ganglia (14).

Sympathetic sensory nerve fibers exit the spinal column through the dorsal root, whereas efferent preganglionic sympathetic axons exit the spinal segments through the ventral nerve roots. These efferent neurons exit the spinal nerve and synapse with postganglionic nerves in the sympathetic ganglionic chain, or they may traverse the paraspinal ganglia and synapse with postganglionic fibers in prevertebral ganglia which regulate visceral activity (18).

Aside from the anatomic segmentation of neuroanatomy, a second necessary assumption for thermography, derived from embryology, is the anatomic symmetry of the human body. This assumption is, of course, not valid for visceral organs, which may have asymmetric distribution within the thoracic and abdominal cavities, but it is evident for somatic structures in an otherwise normal person. The right to left symmetry of the skeleton, musculature, skin, nervous system, and somatic vascular tree permits the assessment of heat loss from symmetric body surfaces as a reflection of physiological homeostasis. Since these tissues are symmetric, any asymmetry of thermal pattern may reflect an alteration in regulatory processes secondary to pathophysiological insult (19–21).

PHYSIOLOGICAL CONSIDERATIONS

Regulation of the cutaneous blood flow and surface heat radiation is generally considered a function of the autonomic nervous system. Most somatic vasomotor control is regulated by the sympathetic branch. As indicated above, preganglionic sympathetic motor neurons exit the spinal canal through ventral nerve roots and synapse with postganglionic sympathetic neurons in the paraspinal ganglionic chain, using acetylcholine (Ach) as neurotransmitter (Fig. 5.1). Acetylcholine receptors on postsynaptic membranes are either nicotinic or muscarinic (based on their inhibition by these chemicals). Most postganglionic vasomotor neurons have nicotinic receptors, whereas most other sympathetic fibers, as well as smooth muscles, have muscarinic receptors. Preganglionic sympathetic axon branches may pass to nodes both rostral and caudal to the segment where the fiber exited. The postganglionic fibers with which they synapse innervate vascular smooth muscles using norepinephrine (NE) as neurotransmitter. This neural networking produces an area of sympathetic, vascular innervation which is larger than the embryonic segment with which it was initially identified; it also produces an overlap of sympathetic vascular control in which sympathetic neurological control is exerted from one or several spinal segments (11). Many postganglionic sympathetic nerve fibers recombine with the spinal nerves and constitute approximately 8–10% of all mixed-function nerves to peripheral somatic structures.

Injury to any peripheral nerve produces both pain and altered sympathetic vasomotor activity. Complete lesion causes cessation of sympathetic tone to vascular smooth muscles; the area of innervation becomes hyperthermic, with increased core temperature blood flowing through the region. Eventually, spontaneous myogenic activity and increased sensitivity to circulating NE causes the region to vasoconstrict and become hypothermic. Irritation rather than section of the sensory or sympathetic fibers generally increases sympathetic activity and causes vasoconstriction, although the immediate response may be vasodilation, using a mechanism described below.

Afferent sensory nerve fibers in the dorsal root transmit the gamut of sensory information. A-Δ and C-fiber neurons transmit the perception of nociception or pain. These neurons have small diameter axons, most of which are unmyelinated. The axons terminate as branched free nerve endings in somatic structures and synapse with central interneurons in the lateral dorsal horns of the spinal column. Their cell bodies are located in the dorsal root ganglia, imbedded in the intervertebral foramina, where they

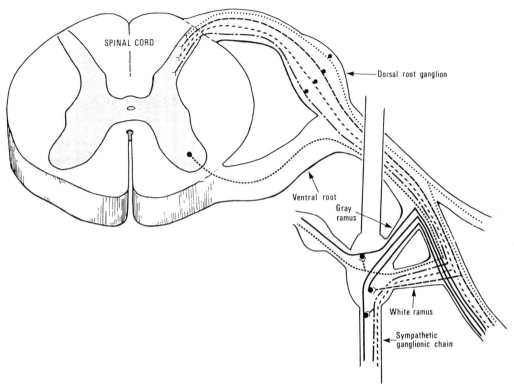

Figure 5.1. Proposed paraspinal sensory and vasomotor nerve pathways. Afferent somatic sensory neurons (·····) may dichotomize, sending axons into different nerves. Somatic sensory nerves may also send branches to synapse with postganglionic sympathetic nerves (-——·), or they may even branch to innervate visceral organs (- - -), thus transmitting both visceral and somatic sensation. Afferent visceral nociceptive fibers (—–—) may send branches to synapse with postganglionic sympathetic nerve cells, producing vasomotor responses in somatic structures. Preganglionic sympathetic fibers (•••••); postganglionic fibers (———).

synthesize neurotransmitters (and neuromodulators?). The major neurotransmitter or neuromodulator produced by C-type nerve cells is substance P (SP) (22). This undecapeptide is transferred by axoplasmic flow from the cell body to the central synapse within the spinal cord, as well as peripherally to dendritic terminals in the skin and muscles (23). Although its function as a CNS neurotransmitter or neuromodulator has been well documented (24), over 70% of the SP synthesized in dorsal root cells follows axoplasmic flow to peripheral terminals (25). Its fate at these peripheral terminals is uncertain. It has been suggested that SP is destroyed within the terminal (26) or possibly released on a continual basis (27), providing a basal level in peripheral tissues. Substance P has been demonstrated to occur in the skin of a large variety of species, and it is most concentrated in highly innervated tissues such as the nose and palmar surfaces of the extremities (28). The suggestion that SP, or another neuromodulator, may act as a tonic antagonist to sympathetic vasoconstrictor tone, although not clearly documented (29), compliments our general physiological concept of homeostatic balance (30). Most systems in constant activation, such as sympathetic vasomotor tone, are also constantly opposed in order to maintain an intermediate steady state.

Sensory stimulation of afferent C fibers produces pain and avoidance responses. Reflex vasomotor responses may follow through activation of the sympathetic system. The response to painful peripheral stimulation may be vasoconstriction or vasodilation. Vasoconstriction from increased sympathetic activity on arteriolar smooth muscle is a well-documented spinal reflex, but considerable evidence suggests that vasodilation from painful stimulii may also occur, involving direct vasodilatory activity in addition to the simpler model of reduced sympathetic activity (31).

CUTANEOUS NEUROVASCULAR REGULATION

Substance P has been shown to inhibit Ach activation of postganglionic sympathetic nerves; in fact, C fibers containing SP have been found within certain sympathetic ganglia (14). The function of SP within these sympathetic ganglia is unclear, but it has been shown to inhibit Ach at nicotinic receptor sites. Consequently, its release in sympathetic ganglia may inhibit postganglionic activity of vasomotor neurons and produce vasodilation by another mechanism (32).

Substance P has also been shown to have several activities other than as a CNS neurotransmitter, although the physiological significance of those activities remains in dispute. Substance P is a potent vasodilator, both directly and through its influence on mast cell histamine release (18). Antidromic stimulation of C-fiber axons produces increased SP in the tissues and local vasodilation. This C-fiber-mediated, antidromic, vasodilatory activity is probably the mechanism for the "axon reflex" described by Lewis (33) and Krogh (29). The correlation of other findings suggests an even more significant role for SP and sensory C fibers. Cuello et al. (14) found that C nociceptive neurons containing SP sometimes bifurcate and send branches to synapse with sympathetic postganglionic fibers, thus describing another extraspinal reflex arc in which sympathetic control is exerted through synapses in the sympathetic ganglia rather than through CNS interneurons. Substance P has been demonstrated to inhibit nicotinic Ach receptors on postsynaptic nerve cells (32), resulting in peripheral vasodilation.

It has been argued that nociceptive C-fibers in the dorsal root are normally quiescent, with little background discharge, but that stimulation of the dorsal root or ganglion produces both centrally and peripherally directed impulses. These orthodromic and antidromic impulses produce the perception of pain and cause vasodilation and hypothermia in the distal cutaneous tissues, presumably through the axon reflex release of SP from the dendritic terminals. Acute compression of the dorsal nerve root results in a burst of impulse activity, rapidly adapting with a return to normal, but similar acute compression of the dorsal root ganglion produces extended periods of neuron firing (34). Furthermore, chronic injury to either the dorsal nerve root or peripheral nerve increases the sensitivity to acute compression, such that long periods of repetitive firing, lasting up to 25 minutes, may be initiated by mild acute compression of dorsal roots of injured axons.

Wall and Devor (27) have similarly demonstrated a low tonic level of spontaneous afferent neuronal activity in the dorsal root ganglion of rats. These
impulses proceeded both orthodromically and antidromically and were increased by mild compression of the dorsal root. Finally, trauma or chronic
irritation of afferent C-fibers ultimately results in depletion of SP from both
ends of the axon and eventual cessation of SP production by the dorsal root
ganglion cells (35).

THERMOGRAPHIC IMPLICATIONS

The results just described suggest an overall model that could explain the
thermal image of paraspinal hyperthermia as well as peripheral, dermatomal
hypothermia associated with chronic spinal pathology. Wall and Devor (27)
have suggested that the spontaneous activity of dorsal root ganglion cells may
have a physiological function, perhaps by causing peripheral release of neuro-
peptides (SP?). If this were true, it would satisfy the same suggestion by
others who have described efferent vasodilatory activity on sensory nerve
stimulation (29). Moreover, spinal trauma may sensitize C nociceptive fibers
for increased antidromic activity caused by mild compressive irritation, resulting in enhanced peripheral vasodilation and hypothermia. Continued
dorsal nerve root irritation, on the other hand, would lead to eventual depletion of SP from C nociceptive neurons and cell degeneration. The final result
would be chronic vasoconstriction from loss of tonic vasodilatory SP release.

Other anatomical and physiological considerations are also gaining recognition as possible contributors to the thermographic image. Langsford and
Coggeshall (16), for example, have demonstrated that dorsal root ganglion
cells may possess two or more axon branches in the peripheral tissue. Other
researchers have supported this finding (17), indicating that at least 2.3
axons exist for each dorsal ganglionic cell. These dichotomizing sensory
axons may innervate contiguous somatic areas, or they may innervate discrete tissue locations. In either case, their existence suggests one explanation
for the observation that the sensation of pain from a single discrete stimulus
may be perceived as originating in separate areas of the body. In fact, pain
and associated physiological responses may be transmitted in somatic tissues through an axon reflex originating near the spinal column (27).

The axon reflex was first described by Lewis (33) as a nerve phenomenon in
which sensory nerve depolarization was transmitted orthodromically from
the site of origin toward the CNS, while concurrently being transmitted
antidromically toward the periphery, along branches of the same neuron
carrying the original message. The axon reflex was first postulated to account
for the finding of local inflammation and pain in areas adjacent to tissue
injury. Called the "triple response of Lewis," the reflex was thought to be
caused by the release of a substance that stimulated mast cell release of
histamine. This chemical has been variously identified as adenosine tri-
phosphate, histamine, prostaglandin, bradykinin, or SP (36), but recent
evidence strongly suggests that SP is the primary force that stimulates mast
cells and directly inhibits vascular smooth muscle contraction. This phenomenon, in which afferent axon branches release a substance that produces
secondary vasodilation, erythemia, and pain, was generally considered

strictly a local response. The findings that sensory axons frequently branch to different areas casts new significance on the role of the reflex in thermographic interpretation (31,37,38).

Recent studies have further demonstrated that dichotomiting axon branches may migrate to very different organic sites. Pierau, Mizutani, and Taylor (17) found that in certain species dichotomizing axons are located in mixed function nerves to both visceral and somatic structures. In rats they found branches from the same cell in such nerves as the genitofemoralis and saphenous, the coxygeal and pudendal, the tibial and peroneal, and the sural and peroneal. Cell branches were also found in the pudendal and sciatic, tibial and sural, and splanchnic and intercostal nerves. Similar dichotomizing fibers have been demonstrated in cats, using cold-induced vasodilation and double-labelling techniques in both nerves. These findings have significant implications for our understanding of somatovisceral, somatosomatic, and viscerosomatic reflexes. Axon branches in both the splanchnic and intercostal nerves, for example, may help to explain the existence of "referred pain," in which pathology of certain visceral organs often produces pain at somatic sites uniquely associated with that organ. Gallbladder disease frequently produces pain in the muscle tissue of the right subscapular region. The splanchnic nerve innervates the gallbladder, whereas subscapular innervation occurs through the intercostal nerve. Similar arguments, based on neuroanatomic mapping of dichotomizing sensory axons, may contribute to our understanding of pain referral and thermal findings in appendicitis (6) and myofascial irritation (7).

Purely sensory neurons may bifurcate and send axons to different somatic areas. The termination of the neurons in arbors of small free nerve endings and in close association with cutaneous blood vessels is well documented (7). Afferent, visceral nociceptive C-fiber axons travelling within the sympathetic chain and having branches to sympathetic postganglionic cells have also been found. These cells may influence visceral and somatic vasomotor activity through release of SP from the collateral branches onto postganglionic cell bodies within sympathetic ganglia. Substance P, by inhibiting postganglionic depolarization, could produce vasodilation in the area of vasomotor innervation. Dichotomizing somatic sensory axons, with collateral branches to sympathetic ganglia, could also produce vasodilation through reflex sympathetic inhibition. Dorsal ganglionic cells with dichotomizing axons to both visceral and somatic structures could on the other hand, be responsible for direct referred pain and vasodilation through the axon reflex and release of SP. Continued stimulation of these afferent C-fiber neurons, whether at the root or peripherally, could eventually lead to depletion of SP production and release, resulting in an alteration of pain sensations as well as peripheral vasoconstriction and hypothermia.

THERMOGRAPHIC CONSIDERATIONS

The thermographic importance of these neurophysiological concepts relates particularly to spinal trauma, one of the most important and controversial areas of thermographic diagnosis. Spinal nerve root irritation is difficult to diagnose. Although recent use of computerized tomography and magnetic resonance imaging have improved anatomical imaging of intervertebral disk lesions, pain and physiological impairment are not documented using these technologies. Thermography provides an imaging modality that depicts physiological changes and is highly correlated with pain, both empirically and theoretically.

Afferent C fibers, originating in the dorsal root ganglion, have been found to be directly associated with cutaneous blood vessels and as sensory free nerve endings. These small diameter fibers are anatomically located at the periphery of the dorsal nerve root, and the cell bodies are highly susceptible to compressive irritation (27). Acute irritation may produce an extended burst of activity lasting several minutes (34). Depolarization of these unmyelinated fibers anywhere along the neuron produces action potentials that migrate centrally and peripherally in the axon. As a result, nerve root irritation may produce an extended burst of activity in the dorsal horn of the spinal column as well as in the peripheral receptive field of that spinal segment. Since these nociceptive C fibers contain and release SP both centrally and peripherally, it may be expected that there would be a sensation of pain in the receptive field, as well as vasodilation triggered directly by a release of SP. Continued stimulation of dorsal spinal nerve roots will lead to a decrease in SP production by the ganglion cells (35), and thus a decrease in delivery to peripheral as well as central terminals.

These data suggest that spinal irritation of dorsal nerve roots could produce an immediate vasodilation and hyperthermia in the peripheral area of innervation, specifically the anatomic "dermatome" of sensory distribution. Continued stimulation of the dorsal nerve root could lead to decreased SP synthesis and depletion from distal terminals, resulting in loss of vasodilatory activity followed by vasoconstriction and hypothermia.

The problem with this argument is that it is based on knowledge of basic mechanisms studied in animals rather than humans. We have no proof of their existence in humans. The more generally accepted mechanisms that explained thermographic patterns associated with spinal trauma were first described by Wexler (39), who based his findings on anatomical studies by Pedersen et al. (40). Wexler suggested that the recurrent meningeal (sinuvertebral) nerve, which innervates the dura, vasculature, and ligaments of the spinal column and carries both sensory and sympathetic fibers, is responsible for the thermal patterns seen paraspinally and peripherally. The suggested mechanism involves sensory input from traumatized spinal structures, followed by reflex muscular activity in the associated erector spinae muscles, causing continued irritation and further trauma (37). This positive feedback loop may cause further irritation of the dorsal nerve root, sensation of pain from peripheral receptors, and sympathetic vasoconstriction in response. The result is classic paraspinal hyperthermia with peripheral hypothermia. This hypothetical mechanism has been challenged as too simplistic (37,41). The reason is that each recurrent meningeal branch innervates several spinal segments above and below its origin from the spinal nerve, whereas the peripheral thermal pattern that is frequently seen follows very closely the sensory dermatomal pattern. Hamilton's suggestion, however, that similar thermal findings should be seen paraspinally and peripherally seems unjustified given Wexler's description which includes different mechanisms paraspinally and peripherally.

The problem still remains that although these mechanisms are well documented in animals, little of this model has been tested on humans. Christiansen (42) was the first to recognize the possible involvement of SP and the axon reflex in the production of a thermographic image. Even though extrapolations to humans based on animal studies have yet to be achieved, several authors have incorporated the concept into our understanding of thermographic physiology (6,37,38). Ochoa (31) has even suggested a clinical syndrome based on antidromic SP release from peripheral nerve endings. It seems certain that all the above mechanisms must play some role in thermoregulation, but which is primarily responsible for the image seen on a given subject is speculative at best. Future studies, especially in a surgical setting, using direct nerve root stimulation with and without sympathetic block may provide essential confirmation of these hypothetical mechanisms.

THERMOGRAPHIC ENIGMAS

Thermography is the diagnostic method of choice for the clinical syndrome variously identified as casualgia, Sudek's syndrome, sympathetically maintained pain, or reflex sympathetic dystrophy syndrome (20,21). This condition is characterized by intractable, excruciating pain, especially in the extremities. Its onset usually coincides with peripheral nerve injury, but the trauma may have been dramatic, such as gunshot wound, or so mild as to have gone unnoticed by the patient. Several hypotheses have been offered to explain the neural mechanisms involved, but whichever is accepted, thermography remains the only diagnostic modality that can document the existence of physiological injury in the early stages.

The earliest recognition of causalgia, which occurred during the Civil War, suggested that a positive feedback loop developed during nerve regeneration following severe peripheral nerve trauma. It was thought that sympathetic neurons developed ephaptic connections with sensory axon sprouts within a neuroma. Tonic neural impulses in sympathetic vasomotor fibers stimulated regenerating sensory neurons, producing the sensation of pain and further vasoconstrictor activity. The resulting positive feedback loop would produce continuous pain, severe vasoconstriction, and ischemia, thus further augmenting the pain.

Although this mechanism may play a role in certain cases of sympathetically maintained pain, it is acknowledged in other theories that several sensory neurons often synapse with a single, wide dynamic range (WDR) interneuron in the dorsal horn of the spinal column. These WDR interneurons summate their input and are activated only when sufficient stimulus is provided. An increase in subthreshold activity from one peripheral afferent neuron may facilitate the WDR to respond to other afferent input more easily. A mild injury could therefore sensitize spinal WDR neurons to respond to otherwise subthreshold sensory input from the periphery. The result, again, would be a reflex increase in sympathetic activity, vasoconstriction, ischemia, and pain exacerbation.

Whichever mechanism may be responsible for reflex sympathetic dystrophy syndrome, only thermography can objectively confirm its existence at a time when only pain, without atrophy, is present. Thermography can confirm a physiological response and objectify the degree of vasomotor involvement. Since only peripheral nerve injury is involved, anatomic imaging is of minimal value, as are doppler and ultrasonic blood-flow studies, which only image blood flow in larger blood vessels.

Another enigma of thermographic imaging concerns the observation that the extremities, especially the digits, of many subjects appear colder than environmental temperature, even after extended periods of equilibration (5). Quantitation of this phenomenon had not been attempted or considered prior to thermographic imaging, but its occurrence has continually plagued thermographers because it makes imaging of these structures extremely difficult. Some have argued that the finding of cold extremities defies the laws of physics (43), since even inanimate objects will equilibrate with environmental temperature. There is, however, a physiological component which is often overlooked.

Hills (44) described the physiological effect of the hemoglobin-oxygen dissociation equation previously described by others (45). He demonstrated that under physiological conditions, hemoglobin may account for the *in vivo* transport of as much as 20% of metabolically produced heat. The equation states that the combining of hemoglobin with oxygen to form oxygenated hemoglobin is an exothermic reaction, releasing approximately 18 Kcal per mole of oxygen bound. Using the same equation, the release of oxygen, like that which occurs in actively metabolizing tissues, is an endothermic reaction in which free hemoglobin absorbs energy (heat) in an activated state. The same fact can be deduced from the hemoglobin-oxygen dissociation; increasing temperature causes a rightward shift in the curve. At higher temperatures a greater number of hemoglobin molecules exist in the energized state, unable to bind oxygen. This heat-dependent unloading of oxygen is functional for most tissues wherein metabolism requires a constant supply of oxygen and heat is produced as a byproduct. In certain vascular beds where little metabolism takes place, however, the release of oxygen from hemoglobin may result in the removal of more heat than is generated from metabolism. This is especially likely if other factors that promote hemoglobin-oxygen dissociation are present. Such things as low pH and high carbon dioxide tension also promote oxygen release from hemoglobin. These conditions are likely to be present in tissues with low blood-flow rates and low metabolism. Under extreme conditions, therefore, oxygenated hemoglobin and low blood flow may act as a "heat pump," absorbing more energy from peripheral tissues than is produced metabolically and "refrigerating" the extremities (46).

Further research is required on many issues involved in thermography. Diagnosis cannot continue to be based on subjective observations of thermal patterns; it must eventually include a physiological rationale for every observation. Experimentation and documentation must start with scientific questions based on sound theories. Thermography has certainly progressed beyond these initial stages. Realistic theories abound, open to confirmation or refutation. Thermographers recognize their lack of complete understanding and continually strive to learn more about this unique form of physiological diagnostic imaging.

References

1. Van J: Thermography–generating heat, but value? *Chicago Tribune* June 9, sec 6, p 1, 1986.
2. Grossklaus R, Bergman KE: Physiology and regulation of body temperature. In Engel JM, Fleschu, Stuttgen E (eds): *Thermological Methods.* Weinheim FRG, VCH Publishers, 1985, pp 11–20.
3. Cooper T, Randall WC, Hertzman AB: Vascular convection of heat from active muscle to overlying skin. *J Appl Physiol* 14:207–211, 1959.
4. Love TJ: Thermography as an indicator of blood perfusion. *Ann NY Acad Sci* 335:429–437, 1980.
5. Hobbins W: Differential diagnosis of pain using thermography. In Ring, EFJ, Phillips B (eds): *Recent Advances in Biomedical Thermology.* New York, Plenum Press, 1984 pp 503–506.
6. Hobbins W: Thermography and pain. In: Gautherie M, Albert E (eds): *Biomedical Thermology.* New York, Alan R. Liss Inc, 1982, pp 361–375.
7. Fischer A, Chang CH: Temperature and pressure threshold measurements in trigger points. *Thermology* 1:212–215, 1986.
8. Travell JG, Simons DG: *Myofascial Pain and Dysfunction.* Baltimore, Williams & Wilkins, 1983.
9. Houdas Y, Ring RFJ: *Human Body Temperature—Its Measurement and Regulation.* New York, Plenum Press, 1982.
10. Keegan JJ, Garrett FD: The segmental distribution of the cutaneous nerves in the limbs of man. *Anat Rec* 102:409–437, 1948.
11. Normell LA Distribution of impaired cutaneous vasomotor and sudomotor function in paraplegic man. *Scand J Clin Invest* 33 suppl 138:25–41, 1974.
12. Hockfelt T, Johansson O, Kellerth JO, Ljungdahl A, Nilsson G, Nygards A, Pernow B. Immunohistochemical distribution of substance P. in: Substance P. U.S. von Euler and B. Pernow eds. New York, Raven Press. pp 117–146, 1977.
13. Bayliss WM: On the origin from the spinal cord of the vasodilator fibers of the hind limb and the nature of these fibers. *J Physiol* 27:173, 1901.
14. Cuello AC, Priestly JV, Matthews MR: Localization of substance P in neuronal pathways. in von Euler US (ed): *Substance P in the Nervous System. Ciba Foundation Symposium* 91:79–99, 1982.
15. Matthews MR, Cuello AC: Substance P immunoreactive peripheral branches of sensory neurons innervate guinea pig sympathetic neurons. *PNAS* 79:1668–72, 1982.
16. Langsford LA, Coggeshall RE: Branching of sensory axons in the peripheral nerves of rats. *J Comp Neurol* 203:745–750, 1981.
17. Pierau FK, Mizutani M, Taylor DCM: Do dichotomizing afferent nerve fibres transmit the axon reflex? in Hales JRS (ed): Thermal Physiology. New York, Raven Press pp 17–20, 1984.
18. Appenzeller O: *The Autonomic Nervous System.* New York, Elsevier Press, 1982.
19. Uematsu S, Edwin DH, Jankel WR, Kozikowski J, Trattner M: Quantification of thermal asymmetry—part 1: normal values and reproducability. *J Neurosurg* 69:552–555, 1988.
20. Uematsu S, Jankel WR, Edwin DH, Kim W, Kozikowski J, Rosenbaum A, Long DM: Quantification of thermal asymmetry—part 2: application in low-back pain and sciatica. *J Neurosurg* 69:556–561, 1988.
21. Newman R, Seres J, Miller E: Liquid crystal thermography in the evaluation of chronic back pain: a comparative study. *Pain* 20:293–305, 1984.
22. Nicoll RA, Schenker C, Leeman SE: Substance P as a transmitter candidate. *Ann Rev Neurosci* pp 227–268, 1980.
23. Brimijoin S, Lundberg JM, Brodin, E, Hokfelt T, Nilsson G: Axonal transport of substance P in the vagus and sciatic nerves of the guinea pig. *Brain Res* 191:443–478, 1980.
24. Otsuka M, Konishi S, Yanagisawa M, Tsunoo A, Akagi H: Role of substance P as a sensory

transmitter in spinal cord and sympathetic ganglia. In von Euler P (ed): *Substance P in the Nervous System. Ciba Foundation Symposium* 91:13–29, 1982.

25. Keen P, Harmar, AJ, Spears F, Winter E: Biosynthesis, axonal transport and turnover of neuronal substance P. In von Euler P (ed): *Substance P in the Nervous System. Ciba Foundation Symposium* 91:145–164, 1982.

26. Quick M, Emson PC: Presynaptic localization of substance P degradative enzyme(s) in rat substantia nigra. *Neurosci Lett* 15:217–222, 1979.

27. Wall PD, Devor M: Sensory afferent impulses originate from dorsal root ganglia as well as from the periphery in normal and nerve injured rats. *Pain* 17:321–339, 1983.

28. Brodin E, Nilsson G: Concentrations of substance P-like immunoreactivity in tissues of dog, rat and mouse. *Acta Physiol Scand* 112:305–312, 1981.

29. Adleman GN, Weiant CW: Photography through the skin proves chiropractic a science. *J Natl Chiro Assoc* April, p20, 1952.

30. Rowell LB: Reflex control of the cutaneous vasculature. *J Invest Dermatol* 69:154–166, 1977.

31. Ochoa, J: The newly recognized painful ABC syndrome: thermographic aspects. *Thermology* 2:65–107, 1986.

32. Ryall RW: Modulation of cholinergic transmission by substance P. in von Euler US (ed): *Substance P in the Nervous System. Ciba Foundation Symposium* 91:267–276, 1982.

33. Lewis T: Experiments relating to cutaneous hyperalgesia and its spread through somatic nerves. *Clin Sci* 2:373–423, 1935.

34. Howe JF, Loeser JD, Calvin WH: Mechanosensitivity of dorsal root ganglia and chronically injured axons: a physiological basis for the radicular pain of nerve root compression. *Pain* 3:25–41, 19 .

35. Jessell T, Tsunoo A, Kanazawa I, Otsuka M: Substance P: depletion in the dorsal horn of rat spinal cord after section of the periphral processes of primary sensory neurons. *Brain Res* 168:247–259, 1979.

36. Burnstock G: Autonomic neuroeffector junctions—reflex vasodilation of the skin. *J Invest Dermatol* 69:47–57, 1977.

37. Hamilton BL: An overview of proposed mechanisms underlying thermal dysfunction. *Thermology* 1:81–87, 1985.

38. Schnitzlein HN: The neuroanatomy and physiology related to thermography. *Academy of Neuromuscular Thermography Proceedings.* New York, McGraw-Hill, 1985, pp 21–25.

39. Wexler CE: *Atlas of Lumbar Thermographic Patterns.* Tarzana, CA, Thermographic Services, Inc, 1983.

40. Pedersen RL, Blunck CFJ, Gardner E: The anatomy of lumbosacral posterior rami and meningeal branches of spinal nerves (sinu-vertebral erves). *J Bone Joint Surg* 38:377–394, 1956.

41. Ash CJ, Shealy CN, Young PA, Van Beaumont W: Thermography and the sensory dermatome. *Skeletal Radiol* 15:40–46, 1986.

42. Christiansen J: Thermographic physiology. In Rein H (ed): *The Primer on Thermography.* Sarasota, FL, H Rein, pp 7–14.

43. Clark RR, Endholm OG: *Man and His Thermal Environment.* London, Edward Arnold Publishers, 1985.

44. Hills BA: Chemical facilitation of thermal conduction in physiological systems. *Science* 182:823–825, 1973.

45. Roughton FJW: The oxygen equilibrium of mammalian hemoglobin. *J Gen Physiol* 49:105–124, 1965.

46. Christiansen J, Vlasuk S: Cold fingers and toes. *Initial* (newsletter) 9:13, 1988.

6

Insurance and Workers' Compensation

Thomas J. Clay, DC, DABCT

Once you begin to use thermography in your practice, you will encounter increased involvement with insurance companies and the insurance departments of self-insured companies. In addition to your expertise as a physician, you will be required to deal effectively with the medicolegal issues of practice and should prepare yourself accordingly. Litigation is further discussed in the following chapter.

Most insurance companies will acknowledge a billing for thermographic studies based on certain established criteria, which may vary from one company and state to another. The criteria may include,

1. Documentation of the medical necessity and appropriateness of the thermographic studies;
2. Evaluation of the fees charged as being "usual and customary" for the region or area in which the test was conducted;
3. Qualifications of the technicians and doctors who performed the studies;
4. Quality of the thermograms;
5. Adherence to standard protocol during thermographic testing.

It has been my experience that reimbursement for fees will depend on how well you communicate with and educate your insurance agent about thermography. Putting myself in the position of the claims agent, I have become more tolerant of the challenges facing both of us in this new arena. From the agent or insurance company's perspective, it makes sense to me that the legitimacy of this thing called "thermography" is questioned. It makes sense that insurers are concerned about whether or not the person performing the test has legitimate credentials. What is a "credential" in the field of thermography, anyway? Where do they go for answers? Isn't it natural to question something new, especially when there is suddenly a mountain of thermographic insurance claims to consider? It makes sense that insurers would want to know if the information provided by a thermographic test could have been obtained as effectively from some other test at less cost. And, how do they determine what the appropriate fees are for such a test?

So, the first thing to do is lower your defenses, tuck in your ego, and get on the phone to the insurance agent. Cooperation, openness, and, above all, information, will do much elicit a favorable reaction to thermography—and a sympathetic attitude toward your bill.

In my clinic, we send a packet of information on thermography with the bill. The first item is an abstract of a 3500-word report prepared for and approved by the American Medical Association's (AMA) Council on Scientific Affairs and released in 1987 (1). The report, one of the most thorough, accurate assessments of thermography to date, was recently referred to by AMA's Office of the General Counsel as "a 'state of the art' scientific report on the safety and efficacy of the use of infrared thermography as a diagnostic adjunctive procedure in the diagnosis of selected neurological and musculoskeletal conditions."

Having established the legitimacy of thermography with this the abstract of AMA's report, you might want to include in your information packet copies of any certification credentials you may have earned, as well as a document listing standard testing protocol and a statement that you have followed these rules and guidelines for all tests performed in your office. The insurer's concern about fees could be resolved or at least minimized by including references from various publications that discuss the average fees charged for thermographic services. You should avoid sending any documents that list "usual and customary fees" in your geographic area, because it could be construed as setting fees in collusion with other thermologists and labs, which is illegal.

You may also find it helpful to include a copy of your thermographic report on your patient in which you establish medical necessity. A guideline for "medical necessity" and "appropriateness" is suggested below.

Necessity

A thermographic evaluation of a patient is considered medically necessary whenever it is the opinion of the treating doctor that the diagnostic information provided by the study will be of use in determining the appropriate treatment plan for the patient, assessing the amount and type of residual dysfunction at the time of examination, or aiding in the diagnosis of the clinical and subclinical status of the patient.

When a second opinion is requested, thermography is considered medically necessary whenever the examining doctor believes the diagnostic information provided by the study is needed to properly assess the patient's current level of dysfunction.

If thermography is used as a basis for determining the necessity or appropriateness of treatment, then it would be appropriate to use one or more follow-up thermographic studies as a basis for modifying or discontinuing the treatment, as well as for helping to determine the extent to which maximum medical improvement can be achieved and the degree of residual dysfunction at that time.

Appropriateness

Thermography provides clinical, physiological diagnostic information not easily obtainable and in most cases significantly different from that obtained by other diagnostic techniques. Consequently, the test is appropriate in all phases of patient evaluation.

Review

Medical thermography is a sophisticated technology which requires appropriate education, training, and skill. Thermographic studies should be reviewed by only those doctors who hold such credentials. Anyone else would not be sufficiently competent to make valid judgments or comments concerning the appropriateness, necessity, or accuracy of thermographic studies.

In many states, including the one in which I practice, it is extremely helpful for the doctor to personally contact the claims agent, especially in cases involving workers' compensation. Because insurance agents are responsible for approving bills for payment associated with your patient's claim, they tend to be intensely interested in what you have to say. I have found from experience that any honest, sincere help you can provide to establish the legitimacy and the exact nature of the injury and a treatment plan, as well as a progress report on the patient's response to treatment and your prognosis, will greatly increase your chances of getting reimbursed for your thermographic and chiropractic services. Moreover, should your patient experience a future on-the-job injury, who would appreciate your release thermograms more than the involved insurance carrier looking for pre-existing injuries?

Reference

1. AMA Council on Scientific Affairs: *Thermography in Neurological and Musculoskeletal Conditions.* Chicago: AMA, Office of the General Counsel, 1987.

7

Litigation

Thomas J. Clay, DC, DABCT

Medical thermographic evidence increasingly is being presented and accepted in trial courts at every level of the federal, state, and local judicial systems of most states. Thermography records the body's surface temperature changes, reflecting cutaneous vascular response to injured pain fibers, as well as those heat changes relative to other irritation or damage that alters blood flow. It has, therefore, become invaluable in the objective assessment of pain. Prior to thermography, pain was considered a subjective complaint which could not be objectively supported.

Thermography is the most significant contribution to bodily injury litigation in the last twenty years, because of its ability to objectify soft-tissue injury and either sensory or sympathetic nerve irritation as well as damage at root, peripheral, and sympathetic levels. It is no longer necessary for the treating physician to be in a position of having to defend his testimony in response to such questions as "Doctor, isn't this only a subjective complaint?" or, "Is this not a 'functional overlay'?" or, "Doctor, isn't your patient over-reacting?" With a positive thermogram, which objectively illustrates the existence of sensory or sympathetic nerve fiber irritation that is consistent with the patient's complaints and correlates with other clinical findings, both the doctor and the attorney representing an injured claimant may enter the courtroom with increased confidence. On the other hand, thermography is also invaluable to the attorney defending against an injured claimant when thermographic evidence is negative and effectively supports malingering.

The material in this chapter is an accumulation of my experience and extensive exposure to depositions and trial testimony. It is my sincere desire to guide you around the "pitfalls" I fell into and struggled out of no more than one occasion. Even so, the best thermography in the world is only as good as your ability to properly interpret, document, and present evidence in a proper legal format that will help a judge or jury understand the patient's injuries and permanent impairments. The following explanation describes how I work. It is by no means the only way, but I have yet to be denied the use of thermographic evidence in a courtroom. It works for me, and it can work for you, too.

Training is your most important asset. Proper training is imperative, not only to make sure you know how to produce good quality thermograms and are able to interpret them accurately, but also to give you adequate qualifications and credentials. Thermograms must be performed according to strict protocol. Any "shortcuts" could be disasterous. Opposing attorneys have copies of standard, accepted protocols and will not hesitate to challenge your procedures and credentials. It is essential that you learn the correct way to perform and interpret thermography from an established source that awards credentials, that you learn standard protocol, and that you follow the rules.

In addition to adhering to protocol, do not settle for substandard thermograms. Remember, as in other documented evidence such as x-rays, computed tomographic or CT scans, magnetic resonance imaging (MRI), and so forth, the opposing side will usually have its experts evaluate your evidence. Do not get caught with poor quality thermograms.

Assuming you are well trained, have proper credentials, and are sticking to protocol and producing consistent, high-quality thermograms, what happens when you plan to use this documentation in a medicolegal situation? If there is any doubt in your mind about what you are doing, do not do it! You could embarrass yourself, or you could be responsible for keeping important diagnostic evidence from the jury if you are not fully aware of the ways to have your documents accepted into evidence. You could conceivably set an undesirable precedent which would carry over to other cases, making it more difficult for others to use thermography as evidence. If you find yourself in this situation, there are many expert, certified thermologists in our profession who would be more than willing to help.

Attorneys you work with are seldom familiar with all aspects of thermography, and it is your job to inform them. When the opposing attorney objects to your "narrative" testimony and is supported by the judge, it means you must stick to answering specific questions addressed to you. Before you can offer testimony about thermographic evidence at a trial, it must be properly introduced. It is indeed embarrassing when the attorney for your side does not know what to ask you. Although it is not your job to properly introduce evidence, you can do much to educate the attorney. This can be achieved in a pretrial conference, during which you can advise counsel on the appropriate questions to ask and provide information concerning the use of thermography in other cases. This opportunity will save both of you humiliation and embarrassment, and do much to facilitate a satisfactory outcome.

During the pretrial conference, it is important that your attorney understand what questions to ask you so that you can establish your qualifications as an expert in front of the jury prior to your testimony. You may be appearing as a thermographic expert who performed the test, or you may be appearing as the treating physician who used the thermographic evidence in the diagnosis and treatment of your patient. The attorney for the other side may attempt to persuade your attorney to "stipulate" your qualifications. Do not let that happen. Doing so would automatically make you accepted by the court as an expert, but you would lose the opportunity to explain the great amount of training and experience you went through to become an expert. The jury should hear your qualifications in as much detail as possible, so they understand that you are more than just a photographer of pretty, colored pictures. Make sure therefore, that your attorney does not "stipulate" your qualifications, and that he or she knows how to ask you about your credentials in such a way that the jury comes to respect your testimony. Remember that during your testimony you are in effect becoming a teacher of twelve captive laypersons who know nothing about you; your credibility as an expert in their minds is paramount.

What follows is a list of questions and answers that I present to my attorney during our pretrial conference; it is in a format that I have found works well in addressing all the issues. This scenario gives me the opportunity to state my credentials, describe my experience, define my area of expertise, present thermography, and introduce my evidence in a professional, calm, and thorough manner, while being asked in a way that is respectful, calm, and professional. This approach tends to disarm and defuse the "other side" before it has a chance to ask some of the same questions in a demeaning tone with distastful theatrics.

Simple language and terminology are used. Your explanations and terminology must be kept as simple as possible because you will be speaking to a jury that comprises a random mix of persons from all walks of life. You cannot expect them to be familiar with your specialty or your terminology, any more than you might be familiar with engineering terms or computer science. Do not be seduced into thinking you will make a better impression by using complicated terms and technical explanations. You are informing the jurors with your testimony, not the attorneys and not the judge. Also remember to direct your testimony to the jurors; look them in the eyes and relate to them as you speak. Again, it is that group of twelve whom you must teach, inform, relate to, and assist in understanding the complexities of the case in question.

Suggested Legal Format for Introducing Medical Thermography

Q1. Doctor, what is your specialty?
   A1. I am a Chiropractic physician.
Q2. What is "Chiropractic?"
   A2. Chiropractic is a drugless, nonsurgical specialty which deals with musculoskeletal problems that effect the nervous system.
Q3. How does one become a Chiropractor?
   A3. By meeting undergraduate requirements sufficient for admission to an accredited Chiropractic college, and completing a course of study equivalent to four years. After earning a Doctor of Chiropractic degree, a graduate must pass a national

board exam as well as the State Chiropractic Board exams before being granted a license to practice.

Q4. Who referred _____ to your office?

A4. (Response).

Q5. What was the purpose of _____'s visit?

A5. To find out what was wrong and what could be done to help him/her.

Q6. Did you take a history and conduct an examination?

A6. Yes, I took a history and conducted an examination (be specific) and performed thermographic studies.

Q7. Thermographic studies? What is "thermography," Doctor?

A7. Thermography is a pain-free, noninvasive medical technique which photographically produces a heat picture, a picture of heat being emitted from the body.

Q8. Doctor, you described thermography as being "noninvasive." Define "noninvasive," please.

A8. Thermography does not require that anything pass through the body such as x-ray, or that needles be put into the body, such as with the EMG or myelogram. Thermography images what is being emitted from the body; it does not put anything into the body to get information.

Q9. Is thermography a chiropractic technique?

A9. No, it is a medical technique, used in the medical field as well as in chiropractic, dentistry, and other areas of specialty.

Q10. What are your qualifications to perform medical thermography?

A10. A. I am certified. (Explain how and by whom) or,

B. I am qualified by training and experience, and I am in the process of achieving certification.

Q11. Doctor, what is your training and experience?

A11. Describe all your training and experience here.

Q12. Have you correlated and compared your thermograms with other established diagnostic methods such as EMG, myelogram, and CT and MRI scans?

A12. Yes, I have. My findings, when compared to EMG, myelogram, CT and MRI scans, and surgical evidence, were at least 95% accurate.

Q13. How do your correlations compare with other published studies?

A13. Dr. Pochaczevsky, professor of radiology at State University of New York and chief of diagnostic radiology, Long Island Jewish Hillside Medical Center, has stated in an article published in the *American Journal of Neuro-Radiology*, " . . . compared with surgical findings, the overall accuracy of contact thermography was 95% and that of myelography 84%." I have found this to be true in my experience also.

Q14. What does medical thermography do?

A14. It objectively assesses sensory or sympathetic nerve fiber irritation or damage, vascular blood-flow changes, and other soft tissue injuries.

Q15. What do you mean by "sensory or sympathetic nerve"?

A15. Much like a telephone cable that contains many smaller wires inside it, each spinal nerve that comes out of the spinal cord contains several different types of nerve fibers. The sensory fibers are one such type, which send pain messages back to the spinal cord and on to the brain. The sympathetic fibers send messages outward and are responsible for functional control of the blood vessels, sweat glands, and so forth. A third type, the motor fibers, are responsible for muscle function and control. Prior to thermography, we could *objectively* assess damage or irritation to the motor fibers only.

Q16. You also said that thermography "objectively" assesses sensory and sympathetic nerve fiber irritation. Would you define "objective?"

A16. Objective, as opposed to subjective, is evidence that can be seen or observed by the examiner without any input from the patient.

Q17. Please define "subjective."

A17. Subjective refers to a symptom as it is experienced by the person feeling the symptom or pain, such as a headache. The patient can say he or she has a headache, but no one else can confirm it. The doctor cannot see it or touch it. The doctor cannot prove it is even there except by the patient's word that the pain is there.

Q18. Could a thermogram then be called a "picture of pain?"

A18. No. The 1987 report of the AMA's Scientific Council states that thermography is not a picture of pain; thermography can help show *why* there could be pain by taking a picture of temperature changes which would indicate whether something is wrong.

Q19. Can thermography prove pain?

A19. One person cannot "prove" another person's pain. Pain is subjective; that is, pain can be felt by the patient only. However, we *can* prove there is irritation or damage to the sensory or sympathetic nerve fibers, or both, which transmit messages of pain. The level of pain experienced depends on the patient's tolerance or threshold. This can vary from day to day and from person to person. A thermogram shows a picture of physiological response that often correlates to pain.

Q20. What is the premise on which thermography is based?

A20. That the heat that comes from the body is generally the same temperature on the right side as it is on the left side. This is considered normal.

Q21. Can there be a "normal" change in temperature from one side to the other?

A21. Yes, in very heavy, muscular individuals, a normal difference in temperature of up to 1°C in the upper forearm is possible. However, this is rare.

Q22. What do you look for when interpreting a thermogram, Doctor?

A22. Changes in the heat patterns when comparing an area of the right side of the body with the same area on the left side of the body. I would like to further explain that by showing several slides.

Q23. How are the heat patterns illustrated?

A23. By different colors, which represent different temperatures.

Q24. Doctor, does a normal (negative) thermogram mean that the patient does not have pain?

A24. No, it does not. The patient may or may not have pain. What it does mean is that the cause of the pain does not appear to be a sufficient irritation or damage to a sensory or sympathetic nerve to cause cutaneous vascular changes.

Q25. Can pain come from other sources and not be detected on a thermogram?

A25. Yes, there are other sources of pain, such as pain which is mental in origin, that does not show up as a change in the heat pattern on a thermogram.

Q26. And what do changes in symmetry indicate?

A26. They indicate there is something wrong.

Q27. Will you please elaborate?

A27. When nerve roots or spinal nerve fibers are irritated, the response to the irritation is muscle spasm along the side of the spine. On a thermogram this shows up as a "hot" spot. This same irritation will follow a dermatome associated with that nerve into the arms, legs, hands, and feet, and will usually show up as cold.

Q28. What is a "dermatome"?

A28. A dermatome is a specific area on the surface of the skin which corresponds to a specific nerve root. For example, there is a definite pattern for the fifth cervical, sixth cervical, and so forth. In fact, we might say that each spinal nerve of the body has a specific dermatome pattern.

Q29. Can a thermogram be faked?

A29. No, not if it is performed according to established rules.

Q30. Please describe the rules, Doctor.

A30. The patient receives a set of instructions prior to the examination, such as not to consume hot or cold beverages for at least one hour before the test, or not to smoke for at least two hours before. Also, there is a pre-exam history during which the patient is asked questions regarding his or her complaints. The patient is asked to mark on a diagram the location of any pain. This helps, during interpretation, to compare the thermography data with the patient's complaints. The patient disrobes and is gowned in such a way that the areas to be examined are exposed. The patient must then wait for his or her skin temperature to become the same as the surrounding temperature in the room before the test is started. He or she is instructed not to rub, scratch, or otherwise touch the areas to be tested. The patient is always attended by a doctor or trained technician to make sure the rules are followed.

Q31. Why is it important that the patient not touch the area to be tested?

A31. This would cause an artifact.

Q32. What is an "artifact?"

A32. An artifact is a hot or cold area on the skin created by touching or rubbing or other contact, which could interfere with the test results.

Q33. Please continue, Doctor. What happens after the patient is prepared for examination?

A33. The photographic views are taken according to the rules, using established imaging positions. Either Polaroid or color 35 mm film is used. The entire exam is repeated a second and third time at 20-minute intervals. This repetition is done to establish consistency and to rule out artifacts. If any artifacts are present, they will disappear on the follow-up sets of thermograms. In some cases, a fourth set is taken with the patient stressed. This we call a stress thermogram.

Q34. Why do you do all this?

A34. Quality control is important. Rigid rules must be adhered to so that each patient is examined in the same way. If the same patient were to return, everything would be done in exactly the same way so that the data from one set of pictures can be compared to the later set of pictures. We do this to collect the most accurate data possible.

Q35. Why was thermography performed instead of an EMG, myelogram, CT scan, or MRI?

A35. Thermography is the only known *objective* method to assess sensory and sympathetic nerve irritation or damage. The other methods do not evaluate sensory or sympathetic pain nerves. Thermography is also found to be more sensitive.

Q36. Based on thermograms, Doctor, can you determine the cause of a patient's injury?

A36. No. A thermogram is very specific as to the presence of irritation or damage to sensory or sympathetic nerves. It tells us that "something" is wrong, not "what" is wrong.

Q37. What important questions can thermography help us answer?

A37. It can tell us whether there is something there or not. It can tell us whether the patient is malingering. It can indicate a nerve injury or irritation. It can indicate substantial soft tissue injury. It can tell us if there is a reason for the pain.

Q38. Would you say that the AMA has taken an official stand on thermography?

A38. The AMA Council on Scientific Affairs submitted a report on February 23, 1987, which included this statement: "Thermography is a safe and effective adjunctive physiologic

test used in diagnosis of selective neurological and neuromusculoskeletal conditions. In those applications, thermography does not stand alone as a primary diagnostic tool. It is a
test of physiologic function that may aid in the interpretation
of the significance of information from other tests."

Q39. Has any federal regulatory agency such as the Department of
Health and Human Services of the Public Health Service or the
Food and Drug Administration (FDA) recognized thermography as a
legitimate test?

A39. Yes. Thermography is recognized by the Department of
Health and Human Services, and the FDA has approved all
equipment used in thermographic testing. Medicare also recognizes thermography.

Q40. Has the U.S. Department of Labor approved thermography in
Workers' Compensation cases?

A40. Yes, since June, 1988.

Q41. Does the American Chiropractic Association have a policy on thermography?

A41. Yes. The policy was released June 23, 1988.

Q42. Doctor, do you have any material that would help the jury understand medical thermography?

A42. (Present illustrations.)

Q43. Do you have with you today the thermograms taken of
_____ on _____?

A43. Yes I do.

Q44. Would you please show and explain them to the jury?

A44. (Present thermograms and explain.)

Q45. Are these thermograms of good diagnostic quality?

A45. Yes.

Q46. Are these thermograms normal or abnormal?

A46. Abnormal.

Q47. Do these thermograms demonstrate abnormal physiology due to
nerve fiber irritation or damage?

A47. Yes.

Q48. Please explain.

A48. The thermograms demonstrate asymmetry of the heat emission patterns following various dermatome patterns.

Q49. Do these films illustrate any soft tissue injury?

A49. Yes. They indicate trigger points, facet irritation, vascular lesions, etc.

Q50. Do the thermograms demonstrate a reason for the complaint of
pain?

A50. Yes. The colder asymmetrical pattern results from irritation
or damage to the sensory or sympathetic fibers and is consistent with the patient's complaints.

Q51. Do these thermograms show an abnormal vascular blood flow?

A51. A. Yes. Abnormal changes in microcirculation of the skin (dilation or vasoconstriction).

        B. Yes. Peripheral vascular shut-down (obstructions or vaso-constriction of the peripheral vessels—.

        C. Yes. Dilation of peripheral vessels in inflamatory process and in serious nerve damage, resulting in causalgia.

        D. Yes. Varicosities.

Q52. Doctor, what is your impression of the findings in these thermograms?

    A52. It is an abnormal study. (Explain)

Q53. Was established protocol followed in this examination?

    A53. Yes.

Q54. Doctor, how much did you charge for this thermographic examination?

    A54. (Response)

Q55. Are these fees usual and customary?

    A55. (Response)

Q56. How much will you charge for today's testimony?

    A56. (Response)

When testifying, be prepared. Make sure proper equipment is present in the courtroom to properly display your thermograms. Not all courtrooms have this equipment, and your lawyer usually does not know exactly what you need. Educate him or her, and arrive early to make certain your equipment is present and in good working order.

This information should guide you through a smooth transition into the medicolegal arena. It should also help you develop the ability to professionally and confidently handle a deposition or a courtroom situation. It has been my experience that very few litigation cases reach the courtroom once the physician becomes proficient in demonstrating permanent nerve damage by means of medical thermography. An expert witness armed with such convincing evidence, along with a reputation for poise and preparedness, is too foreboding an adversary to contend with; a settlement is preferred 90% of the time. Whatever the case, you can learn to enjoy the challenge of a medicolegal experience when you have the confidence in your knowledge of what is wrong with your patient and your ability to effectively explain it to others.

Part **II**

Case Studies

8

Thermographic Imaging of a Patient with a Pancoast Tumor

Geoffrey Gerow, DC, DABCO, Michael Poierier, DC,
Robert Gaik, DC, and James Christiansen, PhD

Pancoast tumor is an uncommon tumor located in the superior sulcus of the lung. It was first described in 1838 and characterized by H. K. Pancoast in 1932. Histologically Pancoast tumors are usually caused by epidermoid or adenocarcinomas of the lung. Occasionally they are caused by small cell carcinomas. Clinically patients will typically present with shoulder, arm, and hand pain, neurological involvement primarily of, but not limited to, the distribution of the ulnar nerve, and Horner's syndrome (ipsilateral ptosis, meiosis, and facial anhydrosis). The diagnosis is classically confirmed by plain film radiography; however, according to one study (1), 6% of patients will have normal x-ray findings. Since Pancoast tumors characteristically affect the sympathetic nervous system, thermographic evaluation can play a role in arousing clinical suspicion of this disease.

We describe a case in which thermography was used to characterize sympathetic involvement with a Pancoast tumor. We postulate that thermography can be a useful diagnostic procedure in helping to identify the sympathetic involvement seen with Pancoast tumor. As such, thermography may be useful in early detection and screening of Pancoast tumors, since it is noninvasive and does not expose the patient to ionizing radiation. This may warrant further investigation.

Pancoast tumor is a rare tumor of the superior sulcus of the lung (2). This tumor was originally thought to be "a vestigial fissure formed during the embryonal development of the right upper lobe during migration of the azygous vein" (3), and was otherwise referred to as a superior sulcus tumor. The first description of a superior sulcus tumor occurred in 1838 (4). The patient had been suffering for a month with pain, tingling, and numbness along the left ulnar nerve distribution. A tumor was palpable about the left lower portion of the neck. The left pupil was contracted and the levator palpebrae was nonfunctional. After the patient's death, a postmortem examination revealed a tumor extended into the regions of the brachial plexus, carotid artery, and cervical sympathetic areas.

In studies performed in 1924 and 1932, Pancoast described the clinical features of malignancies of the upper lung apex, including Horner's syndrome, meiosis, ptosis, and facial anhydrosis (5). Most patients who demonstrated the presence of the Pancoast tumor tended to have shoulder or arm pain and had atrophy of the hand muscles (6).

Pancoast tumors have been reported to produce pain of such intensity that patients have been known to burn themselves in an attempt to obtain relief (1). Patients often experience a loss of appetite from the pain (6). Efforts to control pain associated with Pancoast tumor include percutaneous cordotomy, open cordotomy with rhizotomy, phenol block, partial excision of the brachial plexus, laminectomy, transdermal stimulation, and stellate-ganglion block (7). Respiratory symptoms are infrequent, although the major cause of Pancoast tumor is bronchogenic carcinoma (8). Other disorders, such as laryngeal or thyroid carcinoma, Hodgkin's disease, mesotheliomas, metastatic carcinoma, myeloma, actinomycosis, tuberculosis, and rupture of an echinococcal cyst may produce an apical lung mass (9), along with neurinoma and lipoid infiltration in Hand-Schüller-Christian disease (6). Histologically, analyses have demonstrated that the Pancoast tumor is generally a primary carcinoma of the lung (2) and is usually caused by epidermoid or adenocarcinomas of the lung (10). Occasionally this disorder is caused by small cell lung cancer (10). The male to female ratio is approximately 10:1 (1,2), and most patients are between the ages of 50 and 60 on diagnosis (2).

DIAGNOSTIC CRITERIA

To make the diagnosis of Pancoast tumor, Horner's syndrome is not necessary. If Horner's syndrome is present, however, the likelihood of Pancoast is much greater. Blepharoptosis and cormiosis are the common manifestations present in Horner's syndrome, whereas facial anhydrosis is inconstant (11). The levator palpebrae superioris is striated muscle innervated by the oculomotor nerve. It is unaffected in the sympathetic paralysis of Horner's syndrome. Müller's superior tarsal muscle is a small, sympathetically innervated smooth muscle running from the levator palpebrae to the superior edge of the tarsal plate. When Müller's muscle is denervated as in Horner's syndrome, partial blepharoptosis results, in contrast to the the more complete lid ptosis that occurs in the oculomotor palsies. Other disorders that produce ptosis and may confuse the diagnosis are dermatochalasis, myasthenia gravis, blepharitis, Duane's syndrome, and the levator dehiscence-disinsertion syndrome. Depending on the location, postganglionic lesions may contribute little to anhydrosis. With preganglionic lesions, the finding of anhydrosis may not be apparent unless the patient tends toward profuse perspiration (11). Palpebrae excursion is generally measured by first immobilizing the eyebrow region, whereupon the patient is asked to close his or her eyes and then open them. A ruler is aligned with the upper eyelid margin, and the eyelid should normally traverse a range of 12mm or greater (11). Between 15–30% of the normal population may have anisocoria with a pupil difference of 4mm or greater. To better differentiate simple anisocoria from miosis present in Horner's syndrome, the patient's pupil must first be evaluated in dim light. A discrepancy in pupil size suggests a weakness of the dilator muscle on the side of the small pupil. The pupil affected by Horner's syndrome tends to dilate more slowly in dim light than the pupil affected by simple anisocoria. This "dilation lag" is common in the Horner's syndrome pupil and can best be approximated after 5 seconds in darkness (11).

A diagnosis of Pancoast tumor is usually made by x-ray evaluation. In one study (1), however, 6% of the patients had normal chest x-rays. Often this was because the tumor had been obscured by osseous structure. As such, tomograms, or over-penetrated views, tend to be helpful. The researchers suggested that in the presence of positive clinical findings and negative x-rays, tomograms or exploratory thoracotomy be performed to further evaluate this disorder (1). Fiberoptic bronchoscopy and transbronchial biopsies have provided valuable diagnostic information, particularly as it concerns the histology of this disorder (9).

CASE REPORT

The patient, a 45-year-old white man, presented to a chiropractic college clinic with pain in his right shoulder and arm, as well as numbness and tingling in the fourth and fifth digits of his right hand. The pain had begun approximately 1½ years prior to diagnosis and was described as deep, constant, and sharp. The patient had used a heating pad and applications of Ben Gay®, which he said relieved the "muscle spasms." He noticed the pain most after he returned home from work and relaxed in his favorite chair. The patient also noted that the pain persisted after a night's sleep. The patient had been to four other physicians. Treatment had included B complex, Vitamin C, magnesium, ibuprofin, and tolmetin sodium, which provided little relief. Films of the patient's cervical spine and right shoulder were taken and provided no evidence of any tumor 6 months prior to the patient's arrival at this facility. Electrodiagnostic evaluation performed at the same time as the cervical spine series revealed sensory latencies that were prolonged in the right ulnar sensory nerves, suggesting a right brachial plexopathy, thoracic outlet syndrome, or ulnar neuropathy. The patient gave symptoms of decreased perspiration of the right upper extremity, indicating involvement of the cervical sympathetics. Clinical correlation was required for significance.

After physical examination, it was determined that the active range of motion of the cervical spine was within normal limits, with the exception of right lateral flexion, which produced pain in the supraspinatus region at 35 degrees. Passive ranges of motion were within normal limits without apparent pain. The patient's resting blood pressure was 145/90 on the right and left sides. The neck and upper extremity pulses were normal. The patient demonstrated hypesthesia to vibration on the right side metacarpal-phalangeal area of the thumb, when compared with the left. Range of motion of the right shoulder was limited both actively and passively by pain during flexion, extension, and abduction. Range of motion of the left shoulder was within normal limits. The reflexes were within normal limits bilaterally with the exception of the right triceps, which were diminished. All muscular strengths of the right side were considered to be weak and were given a grade 4. Although the pain was aggravated by performance, this did not appear to be the limiting factor in the patient's ability to exert. The shoulder depression test on the right side reproduced pain into the neck and shoulder areas. The patient demonstrated ptosis and myosis of the right eye. A lack of sweating was present in the right upper extremity.

As a complication of his case, the patient demonstrated pruritic and erythemic papular lesions on his arms, legs, and back. This was later thought to be a contact dermatitis from using a jacuzzi. A radiological examination was performed, which revealed an ill-defined, irregular, radio-opaque mass in the right upper lung field, suggestive of Pancoast tumor (Figs. 8.1 and 8.2).

Thermographic evaluation was performed, demonstrating a segmental hypothermia from approximately the level of the T-4 rib, caudally, both on the anterior and posterior aspects of the right thorax (Figs. 8.3 and 8.4). The patient also demonstrated a relative hypothermia of the fourth and fifth digits on the right side (Fig. 8.5).

DISCUSSION

The diagnosis of this patient's case did not occur from the thermographic tests, but from the radiographic evaluation correlated with clinical findings. Thermography has been used to evaluate sympathetic disorders (12–14) and nerve irritation (15–17). Given that thermography is a reflection of cutaneous blood flow monitored by sympathetic control and that Pancoast tumor characteristically affects the sympathetic nervous system, the resultant hypothermias are noteworthy. If one were to select a lesion that might produce this type of physiological involvement of both the hand and thorax, we assert that one would gravitate to a central, near-the-spine lesion as opposed to a peripheral, extremity lesion. Since the levels involved in the hand region appear to affect an ulnar distribution, one might anticipate the lower cervical or upper thoracic level to be affected. The patient had been complaining to various physicians of this disorder for 1 1/2 years; as recently as 6 months prior to the visit at our facility, he had undergone radiographic evaluations of the cervical spine and shoulder regions as well as an electrodiagnostic study. The evaluation at that time apparently was not definitive of Pancoast tumor. The thermography in this case occurred following the radiological determination of Pancoast tumor. Had the thermographic procedure been performed earlier in this patient's case, a segmental hypothermia as depicted here could have been useful in targeting sympathetic involvement. This could have provided a more aggressive approach to the work-up of Pancoast tumor.

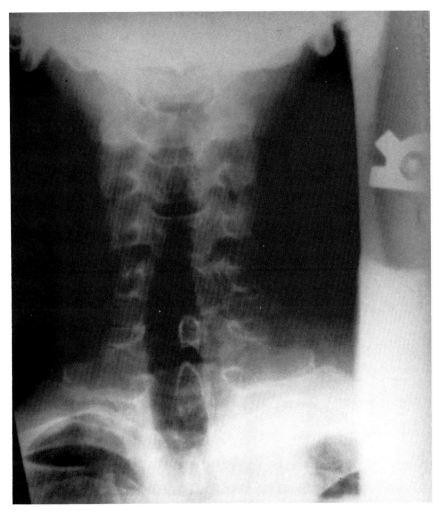

Figure 8.1. Radiograph of cervical spine.

Segmental anhydroses have been reported in other disease states, such as Shy-Drager, multiple sclerosis, pandysautonomia, diabetes mellitus, leprosy, mononeuritis multiplex, and other acute and chronic polyneuropathies (18). It is clear that the survival rate of Pancoast tumor is relatively low. In this case, the patient received exploratory surgery, during which the Pancoast tumor was found to be completely intertwined in the brachial plexus, perhaps accounting for the gross muscular weakness from disuse caused by to pain or neurological embarrassment. The patient was considered terminal and has since died. Characteristically, preirradiation followed by surgical intervention has resulted in a 5-year survival rate of 30–35% (4). Even with this therapy, survival success is grossly limited. Like most tumors, early diagnosis is a prime factor in achieving a successful outcome. Thermography could play a role in the diagnosis of this disorder. More thermographic studies

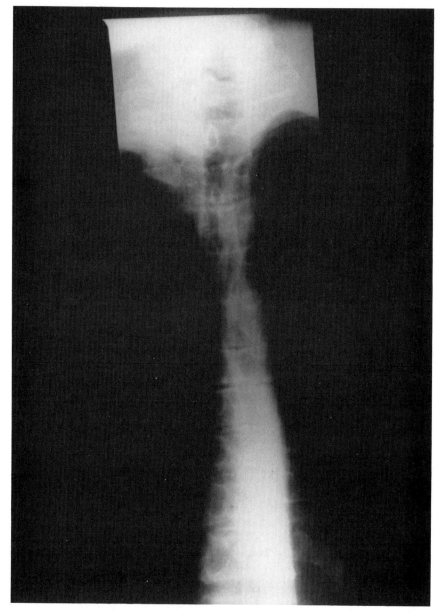

Figure 8.2. Radiograph of upper thoracic spine.

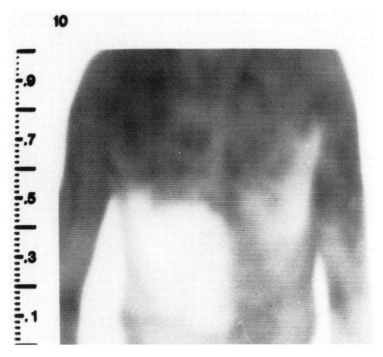

Figure 8.3. Note thermal asymmetry.

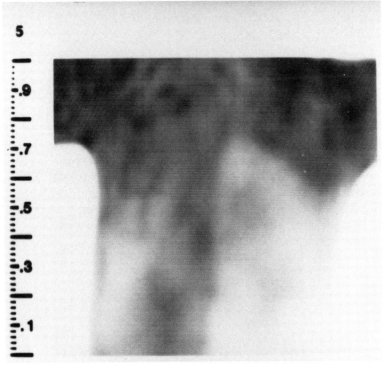

Figure 8.4. Segmental hypothermia is seen.

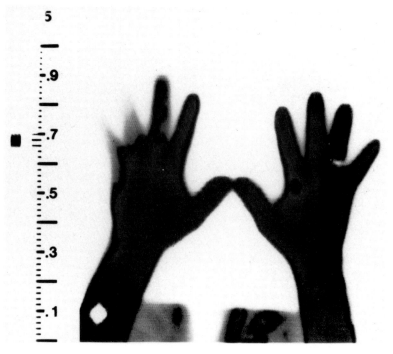

Figure 8.5. The hypothermia extended into the fourth and fifth digits.

on patients who are later diagnosed with this disorder are necessary to further evaluate this assumption.

CONCLUSION

Pancoast tumor is an uncommon tumor of the upper apex of the lung. Characteristically the tumor affects sympathetic nerves and results in Horner's syndrome, a disorder classically characterized by severe shoulder and arm pain. Thermography is a diagnostic modality that is used to evaluate cutaneous blood flow controlled by the sympathetic nervous system, with significant neural networks through the brachial plexus. We have presented a case of Pancoast tumor with associated thermal findings. Had similar thermographic findings been obtained earlier in this disorder, a more aggressive approach to the diagnosis may have resulted in early detection.

References

1. Seidelmann F, Reich N: Pancoast's syndrome. *J Am Osteopath Assoc* 75:126–129, 1975.
2. Kovach S, Huslig E: Shoulder pain and Pancoast tumor: a diagnostic dilemma. *J Manip Physiol Ther* (4):25–30, 1984.
3. Teixeira J: Concerning the Pancoast tumor: what is the superior pulmonary sulcus? (Editorial). *Ann Thorac Surg* 35:577–578, 1983.
4. Mantell B: Superior sulcus (pancoast) tumors: results of radiotherapy. *Br J Dis Chest* 67:315–317, 1973.
5. Krumpe P: Diagnostic approaches to pancoast syndrome. *NY State J Med* 87:320–321, 1987.

6. Mansharamani G, Sunder H, Parvathy S, Krishnamurthy M, Bisht D: Pancoast's syndrome. *J Postgrad Med* 17(2):88–90, 1971.
7. Batzdorf U, Brechner V: Management of pain associated with the Pancoast syndrome. *Am J Surg* 137:638–646, 1979.
8. Paulson D: Carcinomas in the superior pulmonary sulcus. *J Thorac Cardiovas Surg* 20:1095–1104, 1975.
9. Maxfield R, Aranda C: The role of fiberoptic bronchoscopy and transbronchial biopsy in the diagnosis of Pancoast's tumor. *NY State J Med* June:326–329, 1987.
10. Johnson D, Hainesworth J, Greco F: Pancoast's syndrome and small cell lung cancer. Chest 82:602–606, 1982.
11. Thompson B, Corbett J, Kline L, Thompson S: Pseudo-Horner's syndrome. *Arch Neurol* 39:108–111, 1982.
12. Delcour C, Vincent G, DeVaere S, et al.: A predictive thermographic test of the effect of lumbar sympathectomy. *J Mal Vasc* (1):35–37, 1984.
13. Gandhavadi B, Rosen J, Addison R: Autonomic pain: features and methods of assessment. *Postgrad Med* 71:85–90, 1982.
14. Ignacio D, Azer R, Shibuya J, Pavot A: Thermographic monitoring of sympathetic nerve block. *Thermology* 2:21–24, 1986.
15. Dolnitski O, Kuzminskii LN.: Determination of the thermotopography of the hand using a thermograph and liquid crystals in children with injuries of the median and ulnar nerves. *ZH Nevropatol Psikhiatr* 83:1156–1158, 1983.
16. Pochaczevsky R: The value of liquid crystal thermography in the diagnosis of spinal root compression syndromes. *Orthop Clin North Am* 14:271–288, 1983.
17. Pochaczevsky R, Wexler C, Meyers P, et al.: Thermographic study of extremity dermatomes in the diagnosis of spinal root compression syndromes. *Biomed Thermol* 56:339–360, 1982.
18. Faden A, Chan P, Mendoza E: Progressive isolated segmental anhydrosis. *Arch Neurol* 39:172–175, 1982.

9

Thermographic Evaluation in a Patient with Myofascial Pain

Geoffrey Gerow, DC, DABCO, William Pipher, DC, and James Christiansen, PhD

Myofascial pain syndrome and fibromyalgia are important considerations in the evaluation and treatment of chronic pain. Although the two disorders are not mutually exclusive, typical historical, clinical, and thermographic findings can differentiate between them.

Myofascial pain is the most common cause of pain that brings patients to chronic pain treatment centers (1). Myofascial pain syndrome is characterized by trigger points, hyperirritable spots usually located within taut muscule tissue or its associated fascia (2), and pain referral idiosyncratic to the affected muscle or associated fascia (1). The diagnosis of myofascial pain based on physical examination includes an exquisitely tender spot, a band of taut muscle fiber palpably present, lack of full strength through full range of motion in the affected muscle, a local twitch response (a finger run transversely across the muscular fibers produces a muscle twitch), and the pain-referral pattern.

Trigger points are either active or latent. An active trigger point is one that currently exhibits the referral pain pattern, whereas a latent trigger point is quiescent until compressed, whereupon it demonstrates the referral pattern seen in an active trigger point (2).

Myofascial trigger points are sometimes described as myogelosis, myalgic spots, muscle hardenings, muscular non-articular rheumatism, and fibrocytis (2). More recently, fibrocytis (fibromyalgia) has been distinguished from myofascial pain syndromes (3) and is considered a common form of non-articular rheumatism (4).

Fibrocytis is distinct from myofascial pain syndromes in that it has widespread pain with multiple areas of tender points. The patient must present with at least seven tender points to be considered to have fibrocytis (3). The tender points, unlike myofascial trigger points, have no referral pattern, even when compressed. Typically the patient with fibrocytis displays diffuse musculoskeletal aching and stiffness (4) and tends to have morning or general fatigue (3).

The cause of trigger points is not clearly understood, but it is often related to acute muscular strain or repetitive muscular over-use (1). Wide, dynamic range neuron-reflex-arcs, sympathetically mediated, have been proposed as being a factor in the development of trigger points (5). Fibromyalgia syndrome is typically related to a hypoactive person; it has been postulated that a person will experience increased fibromyalgia as he or she ages and does not maintain the length of childhood muscles (6).

Thermography has been useful in identifying myofascial trigger points (1,7,8). Usually the trigger point appears as a well-circumscribed area which is hyperthermic; the referral pain pattern is often hyperthermic as well (9).

Few controlled studies on treatment efficacy for myofascial pain syndrome and fibromyalgia have been performed, and most treatment regimens are based on anecdotal findings. These treatments include local heat, nonsteroidal anti-inflammatory drugs (NSAIDs), biofeedback relaxation training, trigger point injection with local anesthetic, dry needling, acupuncture, electrical stimulation, ischemic compression, and "spray-and-stretch" procedures using vapocoolant sprays.

CASE STUDY

A 51-year-old white woman suffered from a constant, generalized aching pain from approximately the C7/T1 level to L4, as well as an intermittent pain over the lateral aspect of the left middle thoracic rib cage. The patient generally appeared to be tired, in pain, and was continually moaning. The patient explained that she had sought care from a medical physician who had diagnosed her complaint as intercostal neuralgia and provided 25 cortisone injections over a period of 3 months to the inferior and inferior-lateral junction of the breast and chest. Two years earlier, she had had similar complaints and had received 50 cortisone injections to the left breast region over a 4-month period.

The patient's pain was aggravated when sitting in a backless chair and when standing erect and raising both arms above her head. She was more comfortable when sitting erect in a straight-backed chair or when sitting erect followed by laterally bending to the left. The pain decreased while lying supine on a hard surface. The patient had used a heating pad at home, which sometimes relieved the pain enough that she would discontinue her pain medication for a day or two.

The patient's previous medical history included a hysterectomy for ovarian cancer; removal of her gall bladder, benign throat polyps from her vocal cords, and a malignant skin lesion from her forehead; bladder surgery; pneumonia; and a history of peptic ulcers. She had received appropriate care for these disorders and, at the time of her visit for chiropractic care, there were no reported sequelae. The patient was under medication for high blood pressure (labetalol hydrochloride 100 mg tid) and was taking acetaminophen (300 mg) with codeine (60 mg) for pain. The patient denied any previous neurological problems.

Radiographs of the thoracic spine were taken and interpreted by a resident radiologist. The impressions, as shown in Figures 9.1 and 9.2, were thoracic

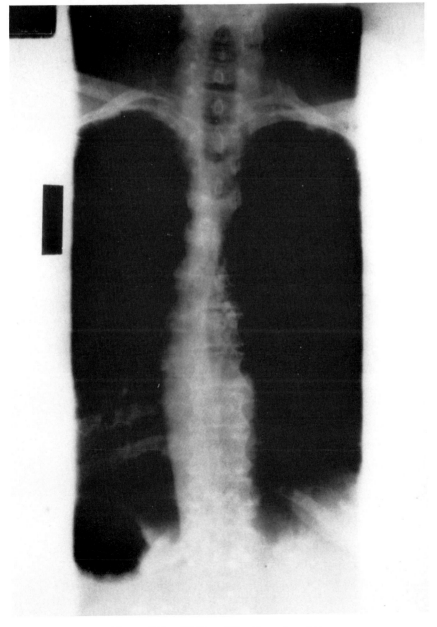

Figure 9.1. AP film of the thoracic spine.

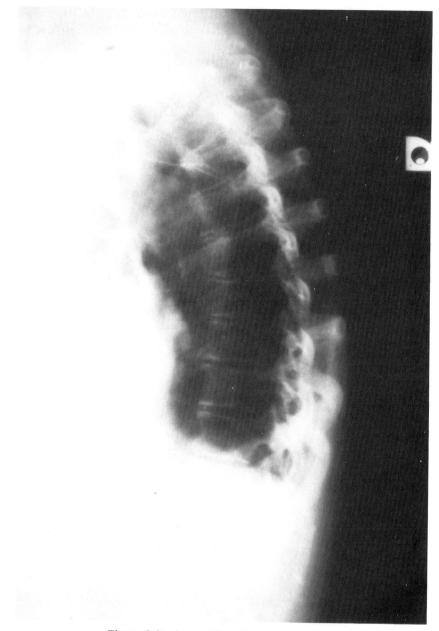

Figure 9.2. Lateral film of the thoracic spine.

spondylosis and mediastinal lymph node calcification, most likely a sequela of a previous inflammatory disease.

Following a routine protocol, a thermographic evaluation was performed. The evaluation revealed three areas of hyperthermicity over the posterior aspect of the back (Fig. 9.3), in the regions of the middle trapezius, upper trapezius, rhomboid, and infraspinatus (Fig. 9.5), and, to a lesser extent, in the latissimus dorsi muscles (Fig. 9.4). Hyperthermia of the left hand in its

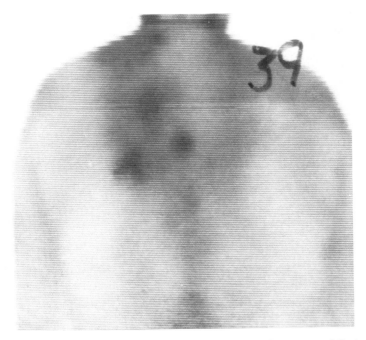

Figure 9.3. Thermogram showing hyperthermicity in three areas of the back.

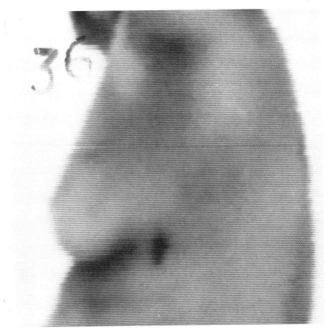

Figure 9.4. Lateral thermogram.

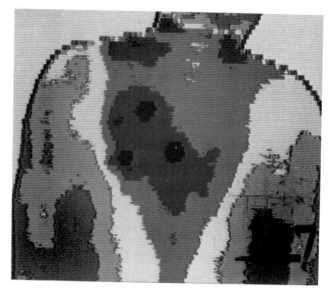

Figure 9.5. Note discrete areas of hyperthermicity (see color plate, page v).

dorsal and palmar surfaces was also noted (Figs. 9.6–9.8, also color plates, page v).

A regional evaluation was performed following the thermographic examination. The patient weighed 160 pounds and was 62 ½ inches in height. Her pulse was 88, respiration 20, and blood pressure 136/80 (right) and 140/82 (left). A breast examination was normal. The patient demonstrated multiple

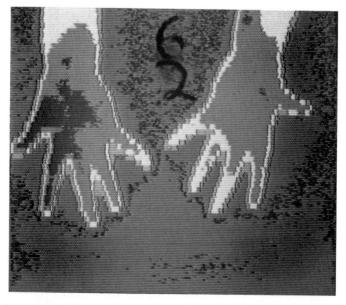

Figure 9.6. Thermal asymmetry in the hands (see color plate, page v).

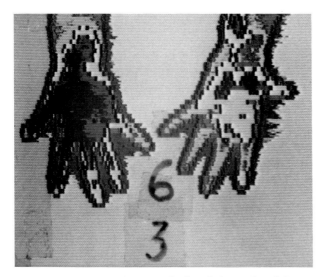

Figure 9.7. Thermal asymmetry in the hands (see color plate, page v).

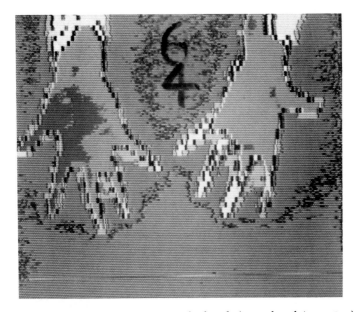

Figure 9.8. Thermal asymmetry in the hands (see color plate, page v).

areas of tender points in the back on palpation. The areas found to be hyperthermic on the thermographic evaluation were marked and pressure was applied over these "hotspots," whereupon pain radiated to the area underneath the left breast. Compression of the spinous processes over T4-T9 were tender to the patient; posterior to anterior compressions and lateral compressions of the chest wall did not, however, aggravate the patient's complaint. Active ranges of motion of the cervical spine and upper extremities were within normal limits, symmetrical, and pain-free. Evaluation of the upper extremity pulses, muscle stretch reflexes, and sensation was normal and symmetrical. Sensory evaluation with a pinwheel of the scapular and torso regions found the left upper trapezius region and left lateral torso relatively hyperesthetic when compared to the opposite side.

DISCUSSION

The thermographic findings of well-circumscribed areas of hyperthermia, which correspond to tender nodular areas that refer pain to the area of the left breast on palpation, are consistent with myofascial trigger points. The hyperthermia noted in the dorsal and palmar surfaces of the left hand may reflect a distal, thermal representation of the myofascial pain syndrome (a). The multiple tender areas in the back did not refer pain on palpation and were consistent with fibrocytis or fibromyalgia.

CONCLUSION

This case demonstrates the possibility that a patient may have two superimposed disorders that are not mutually exclusive. The thermographic data and physical examination suggest myofascial pain in the left mid-back region. The patient's complaints of widespread aching pain and fatigue, together with findings of multiple tender regions without pain referral, suggest fibrocytis or fibromyalgia.

References

1. Simons D: Myofascial pain syndromes: where are we? where are we going? *Arch Phys Med Rehabil* 69:207–212, 1988.
2. Travell J: *Myofascial Pain Syndrome.* Baltimore, MD, Williams & Wilkins, 1987.
3. Wolfe F: Fibrositis, fibromyalgia, and musculoskeletal disease: the current status of the fibrositis syndrome. *Arch Phys Med Rehabil* 69:527–531, 1988.
4. Yunus M, Kalyan-Raman U, Kalyan-Raman K: Primary fibromyalgia syndrome and myofascial pain syndrome: clinical features and muscle pathology. *Arch Phys Med Rehabil* 69:451–454, 1988.
5. Fine P, Milano R, Hare B: The effects of myofascial trigger point injection are naloxone reversible. *Pain* 32:15–20, 1988.
6. Kantor K: Fibromyalgia syndrome. *Postgrad Med* 84:45–46, 1988.
7. Fischer A: Documentation of myofascial trigger points. *Arch Phys Med Rehabil* 69:286–291, 1988.
8. Fischer A: (Letter to the editor). *Pain* 41:411–414, 1987.
9. Silverman H: *Neuromusculoskeletal Thermography*, ed 1. Clayton, GA, Rabun Chiropractic Clinic, Clinical Thermographic Services, 1984.

10

Thermographic Findings in a Patient with Reflex Sympathetic Dystrophy

Geoffrey Gerow, DC, DABCO, Robert Gaik, DC, and James Christiansen, PhD

Thermography has been valuable in diagnosing reflex sympathetic dystrophy (RSD). Following a review of the literature on RSD, we present a case study of a patient afflicted with this disorder in which thermography is used as a diagnostic adjunct.

Reflex sympathetic dystrophy is a disorder marked by a burning pain (1) that is out of proportion to the injury (2). The clinical aspects of this disorder were first described in 1813 (1) and concerned a soldier who had sustained a nerve injury from a musket ball wound of the arm. The soldier's pain was so intense that he requested that his arm be amputated, whereupon he experienced complete relief from the pain.

In 1864 Paget described a glossy skin appearance in a patient who had causalgia. The term causalgia comes from the Greek words "kausis," which means burning, and "algos," which means pain (3). The word "causalgia" was coined to describe the burning pain and dysesthesia associated with this disorder (4). In 1900 Josef Sudeck described bony changes associated with RSD (4).

Although the etiology of RSD is unknown, there are several theories concerning its development. One is the afferent stimuli theory, which asserts that the site of the peripheral injury is where the afferent stimulus begins. This stimulus is mediated by the sympathetic nervous system. Another theory is the internuncial cycle theory, which suggests that the peripheral injury region excites the internuncial pool, increasing motor activity at the distal site and causing a repetitive cycle of irritation affecting the sympathetic nervous supply. A hypothesis that may be related to the internuncial cycle theory assumes that the wide dynamic range, or multireceptive neurons, once sensitized, act to perpetuate painful sensations (5). The artificial synapse theory concludes that at the site of peripheral injury, the sympathetic nerve fibers grow into the afferent nerve fibers, producing a painful stimulus. Another theory, based on the pathogenesis of RSD, describes the result of local inflammation and ischemia, resulting in increased sensitivity of the peripheral vasculature to the norepinephrine. This process produces further ischemia and pain (6).

The disorder is divided into stages. Stage 1 is characterized by pain and sympathetic overactivity (7). Patients have generalized coldness in the affected extremity. Thermography is useful in evaluating such hypothermia. Rarely, the affected limb has been found to be warmer than the unaffected side. Stage 2 is characterized by the symptoms of stage 1 and limitation of movement caused by pain. In Stage 3 both passive and active motion is restricted owing to soft tissue trophic change, which prevents such motion. Stage 4 presents as a fixed, motionless extremity.

Another method of identifying RSD is termed the Betcher's classification, which is defined in order of decreasing severity of the symptoms (2). Grade 1 is the most painful, is not relieved by rest, and results in severe vasomotor and trophic changes and gross restrictions of motion. Grade 2 is characterized by edema, vasomotor and trophic changes in the body part affected, and pain in the region aggravated by movement. Grade 3 is characterized by less pain, edema, and vasomotor changes.

A number of conditions are thought to precipitate RSD, including hemiparesis, myocardial ischemia, cervical osteoarthritis, and discogenic disease (3). Cerebrovascular accident, infection, phlebitis, carcinoma, and neuropathy can also precipitate RSD (6). The greatest precipitating factor is trauma (8).

Typically, the patient's pain totally absorbs his or her interest (1). Placing cool, moist towels over the painful area can sometimes reduce hyperpathic sensitivity.

CASE HISTORY

The patient was a 48-year-old black woman who presented to a chiropractic college clinic with right wrist pain following a Barton fracture. The patient delayed seeking treatment initially, keeping the wrist wrapped in an Ace bandage™ for approximately 3 weeks. After finally seeking care, the patient was treated by inserting a T-bar into the right radial wrist region. She was casted for 6 weeks, on removal of which she experienced decreased range of motion of her wrist and tenderness. The patient was sent to a physiotherapist who used an aggressive approach to passive range of motion, fracturing the patient's ulna. She was then casted for another 6 weeks. Following the removal of this cast, her wrist was painful and stiff. Approximately 7 months after the initial incident, the patient received a diagnosis of reflex sympathetic dystrophy following stellate ganglion block.

The patient sought further care at the chiropractic college clinic, receiving physiological therapeusis to the right wrist area with active ranges of motion and exercise. The patient visited the clinic for approximately 2 years and was then referred to the clinic's department of orthopedics and thermography for examination. On the day of the exam, she complained of pinpoint tenderness in the dorsal wrist region, with a periodic "pulling" sensation in her 5th digit. She described the pain as an ache in the hand and wrist area, as well as a tingling in the wrist, and described her wrist as being very cold.

The patient often felt numbness in her right hand and said the pain caused her sleepless nights. She was able to relieve the pain by soaking her wrist in warm water. Housework of more than 2 hours' duration tended to aggravate her condition. She also described pain in the lateral aspect of her right forearm, as well as in the general elbow region. Her pain was considered constant, i.e., experienced 100% of the time.

The patient received x-ray evaluations, which revealed a comminuted interarticular fracture of the right radius region with little evidence of healing. One month later, following placement of a second internal T-bar, normal healing was observed. A transverse fracture of the distal ulnar metaphysis was noted five months following the comminuted injury of the right radius (Fig. 10.1) At that time, osteopenia consistent with immobilization was demonstrated. The ulnar fracture eventually healed.

One year after the patient's comminuted fracture, films were again taken of her right wrist, which demonstrated osteoporosis of the disuse variety. Four months later, films were retaken, showing that osteoporosis of the right wrist

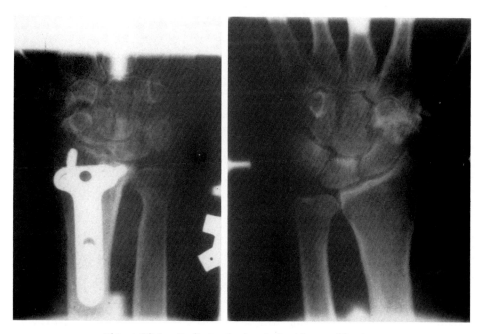

Figure 10.1. Radiograph showing evidence of fracture.

had not progressed significantly, which may have been be related to the patient's therapeutic program.

Thermographic examination revealed hypothermia of the right wrist and forearm; thermal asymmetry was as great as 1.9°C in the hands (Figs. 10.2 and 10.3; see color plate, page v). The hypothermia noted in the right forearm region was approximately 0.6–0.9°C asymmetric relative to the left. The patient's feet were placed in warm water for five minutes. The right hand would not "warm up" relative to the left. Her feet were placed in cold water and slight intensification of the thermal difference was noted.

An orthopaedic evaluation was performed, showing essentially normal vital signs. The patient was right-hand dominant. She demonstrated hyperesthesia with the pinwheel on the right side vis-à-vis the left wrist, and showed a gross weakness in grip strength on the right side. Range of motion, radial and ulnar deviations, and dorsiflexion of the right wrist were limited on the right when compared to the left wrist. Internal and external rotation of the elbow on the right side was also limited vis-à-vis the left wrist.

The patient received active exercise at the clinic, using a wall pulley and performing a range of motion activities.

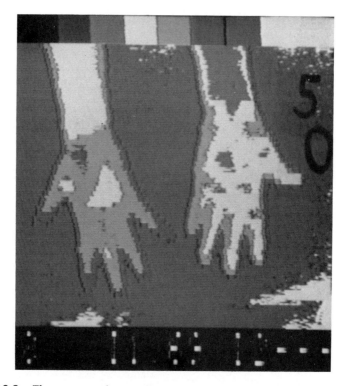

Figure 10.2. Thermogram showing thermal asymmetry in hands (see color plate page v).

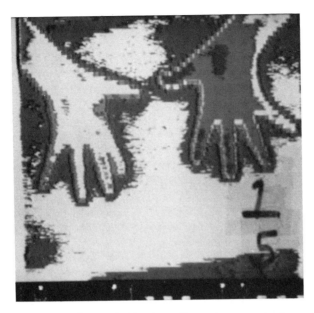

Figure 10.3. Thermogram demonstrating large thermal asymmetry (see color plate page v).

DISCUSSION

This diagnosis of RSD was made through the use of stellate ganglion injection block, which usually requires a local anesthetic. Use of a sympathetic block generally results in reversal of the symptoms, which may improve circulation or may block sympathetic mediated pain (7). Thermography confirmed the clinical picture of RSD (3) and was useful in diagnosing this disorder (9).

The treatment for RSD varies and is based, in part, on the staging of the disorder. In stage 1, early mobilization, physical therapy, and desensitization massage are useful. Stellate ganglion block with ultrasound and ultrasound over peripheral lower extremity nerve fibers are also useful (10). In stage 2, limb elevation following radiant heat, moist heat, paraffin, or fluid therapy may be used. Biofeedback for temperature control has also been beneficial (11). Stellate ganglion block, corticosteroids, and transcutaneous electrical stimulation may produce positive results. Patients with stage 3 or stage 4 disorder may be treated with joint manipulation under Bier block using reserpine or guanethidine (12). Other means of treating RSD include regional intravenous propranolol, unilateral sympathectomy, and thalamic stimulation (10). Contralateral sympathectomies may be used if a unilateral sympathectomy fails therapeutically (13).

Successful treatment of RSD is dependent on early diagnosis (14), which may be perceived by thermographically imaging the fingers and toes of a patient requiring casting procedures (3). Thermography, used in conjunction with the proper therapy, may help prevent this disorder from developing into later stages of disability (14).

CONCLUSION

Presented is a case of RSD diagnosed prior to thermographic evaluation. The thermographic evaluation did, however, corroborate the diagnosis. Early diagnosis of this disorder is crucial for effective treatment and may prevent unnecessary pain and disability for the patient.

References

1. Ecker A: Reflex sympathetic dystrophy thermography in diagnosis: psychiatric considerations. *Psychiatr Ann* 14:787–793, 1984.
2. Watson H, Carlson L: Treatment of reflex sympathetic dystrophy of the hand with an active "stress loading" program. *J Hand Surg* 12:779–785, 1987.
3. Pochaczevsky R: Thermography in post-traumatic pain. *Am J Sports Med* 15:243–250, 1987.
4. Kozin F: Two unique shoulder disorders: adhesive capsulitis and reflex sympathetic dystrophy syndrome. *Postgrad Med* 73:207–216, 1983.
5. Roberts W: Review article: A hypothesis on the physiological basis for causalgia and related pains. *Pain* 24:297–311, 1986.
6. Kobrossi T, Steiman I: Reflex sympathetic dystrophy of the upper extremity: a new diagnostic approach using flexi-therm liquid crystal contact thermography. *J CCA* 30:29–31, 1986.
7. Lewis R, Racz G, Fabian G: Therapeutic approaches to reflex sympathetic dystrophy of the upper extremity. *Clin Issues Reg Anesth* 1(2):1–6, 1985.
8. Subbarao J: Reflex sympathetic dystrophy syndrome of the upper extremity: analysis of total outcome of management of 125 cases. *Arch Phys Med Rehabil* 62:549–554, 1981.
9. Perelman R, Adler D, Humphreys M: Reflex sympathetic dystrophy: electronic thermography as an aid in diagnosis. *Orthop Rev* 16:561/53–566/58, 1987.
10. Portwood M, Lieberman J, Taylor R: Ultrasound treatment of reflex sympathetic dystrophy. *Arch Phys Med Rehabil* 68:116–118, 1987.
11. Barowsky E, Zweig J, Moskowitz J: Thermal biofeedback in the treatment of symptoms associated with reflex sympathetic dystrophy. *J Child Neurol* 2:229–232, 1987.
12. Munn J, Baker W: Recurrent sympathetic dystrophy: successful treatment by contralateral sympathectomy. *Surgery* 102:102–105, 1987.
13. Duncan K, Lewis R, Racz G, Nordyke M: Treatment of upper extremity reflex sympathetic dystrophy with joint stiffness using sympatholytic bier blocks and manipulation. *Clin Orthop* 214:85–92, 1987.
14. Vasudevan S, Myers B: Reflex sympathetic dystrophy syndrome: importance of early diagnosis and appropriate management. *Wis Med J* 84:24–28, 1985.

11

Thermographic Evaluation in a Patient with a Lumbar Disc Herniation

Geoffrey Gerow, DC, DABCO, William Pipher, DC, and James Christiansen, PhD

Back pain affects about 80% of the American population at some point in their lives (1). In a period when disability from chronic low back pain is becoming an important clinical and economic consideration, it is imperative that an early, accurate diagnoses be made in patients with low back pain. Herniation of the intervertebral disc is one etiology of low back pain; it was the primary diagnosis for 14% of patients in a group of 1293 treated for non-specific low back pain (2).

The four primary diagnostic features of lower lumbar disc prolapse are lower extremity radicular pain that approximates a dermatome, signs of neuromeningeal irritation (restricted straight leg raising, well leg raise with reproduction of leg pain), and neurological changes consistent with specific nerve root compression (muscle weakness, sensory change, and hyporeflexia) (3). To confirm clinical suspicions of discal pathology, advanced imaging, such as myelography, computerized tomography (CT), or magnetic resonance imaging (MRI), can be used. However, it is important to correlate imaging and clinical findings. In one study, a population that had never experienced low back pain underwent a lower lumbar spine CT evaluation (4). Nearly one–fifth of this asymptomatic group had disc herniations, and 36% had abnormal CT scans.

Thermography has also been used in the evaluation of low back pain patients. The thermogram is highly sensitive to the presence of discal herniation, being better than 90% sensitive with a specificity of 75%. Typical thermographic findings in patients with low back pain of discal origin include a hyperthermic paraspinal area and stripe of hyperthermicity that may extend beyond the spinal region laterally. This hyperthermic stripe is thought to be associated with irritation of lumbar nerve roots (5). Peripheral hypothermicity may also exist involving affected dermatomal areas.

CASE REPORT

A 35-year-old white man suffered from pain in the posterior left gluteal region, extending down the lateral aspect of the thigh and into the left anterolateral calf. He also described numbness about the dorsum of the left foot. The patient explained that he had attempted to lift eighty pounds of angle iron at work when he felt a snap in his back. He noted no immediate pain but awoke the next morning with low back pain. The patient continued to work for approximately 1 week, after which he sought care from a chiropractic physician. Although his lower back pain remitted after 1 week of therapy, his present lower extremity complaints began approximately 3 weeks after his initial injury.

Following regional examination, pain prevented the patient from assuming the prone position. Straight leg-raising reproduced the patient's symptomatology at 5° of hip flexion when performed on the left and at 15° when performed on the right. Braggard's and Fajerstajzn's tests also reproduced the patient's chief complaint. The double leg raise test reproduced pain into the left lower extremity at 10° of hip flexion. Kemp's test was negative bilaterally, as was the Patrick (Fabere) test.

Thermographic evaluation of the low back and lower extremities was performed with the patient prone, since pain prevented him from maintaining a standing position. Otherwise, standard thermographic protocol was followed. The patient demonstrated a relative hyperthermicity about the lower lumbar spinal region in the general shape of an arrowhead with its apex pointing cephalad (Fig. 11.1), as well as hypothermia about the circumference of the left lower leg. Using the isotherm bar, the patient was seen to have hypothermia of the distal lower extremity on the anterolateral aspect (Figs. 11.2 and 11.3) and about the posterior aspect of the left calf (Figs. 11.4 and 11.5).

Plain film radiographs of the lumbar spine revealed pelvic unleveling, low on the left, with a list of the spine to the right side. Minimum disc wedging was apparent at the fourth lumbar (L4) disc space with an opening on the left (Fig. 11.6). The vertebral structures appeared intact with no definite evidence of osseous or joint pathology. A CT scan was performed, demonstrating a large herniation of the L4 disc with a calcium cap and posterolateral and posteromedial components (Fig. 11.7).

DISCUSSION

This case demonstrates a typical presentation of low back pain caused by discal herniation. Advanced imaging findings, which confirm the presence and location of the herniation, enable us to better discuss thermographic

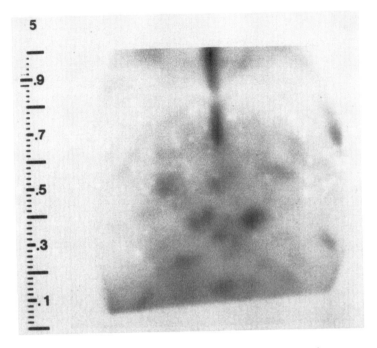

Figure 11.1. "Arrowhead" hyperthermicity in the lower lumbar spine.

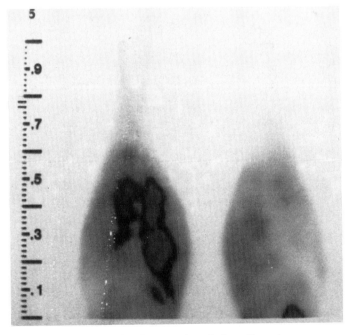

Figure 11.2. Lower extremity hypothermia.

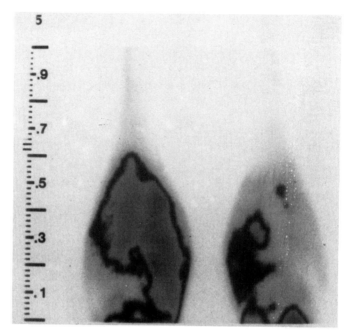

Figure 11.3. Hypothermia on the anterolateral aspect of the lower extremity.

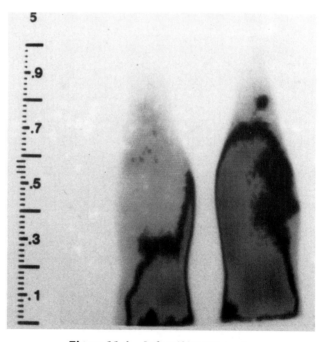

Figure 11.4. Left calf hypothermia.

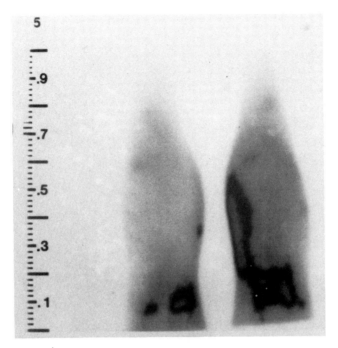

Figure 11.5. Hypothermia on the left lower posterior calf.

abnormalities that may be seen in this type of case. The areas of hypothermicity, as viewed over the circumference of the distal leg, are thought to
encompass the dermatomes of L5, S1, and S2. The thermographic findings
tend to indicate a multilevel nerve embarrassment. It is thought that the
dermatomal hypothermicity may be related to substance P production or
alternation.

Substance P is produced in the dorsal root ganglion and flows down the
nerve by axoplasmic flow to inhabit the entirety of the nerve. It is a neurotransmitter, has vasodilative properties, and is liberated in response to C-
fiber activity initiated by pain. There is thought to be a balance between
normal noxious stimuli and C-fiber activity, producing vasodilative responses and the sympathetic tonus present in the area of involvement. In the
presence of chronic pain, however, substance P stores within the peripheral
nerve are used up; the peripheral nervous system is depleted of its vasodilative properties and comes under sympathetic control, resulting in a cold
extremity. This theory may account for the dermatomal thermal change
described by other thermographers relative to discal involvement. As such, a
positive thermographic evaluation tends to suggest further imaging, such as
CT or MRI, of the lumbar spine to visualize the presence and extent of the
lesion.

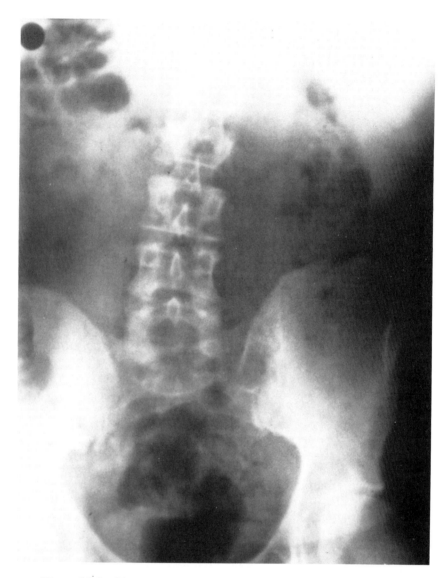

Figure 11.6. Minimal L4 disc wedging noted on plain film radiography.

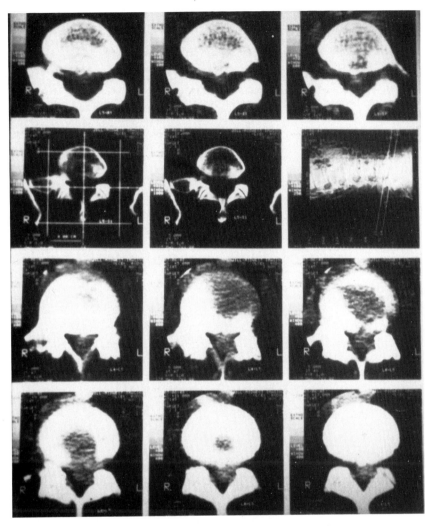

Figure 11.7. CT scan demonstrating disc prolapse.

CONCLUSION

The lumbar disc syndrome is a relatively common finding in a chiropractic office. The thermogram, though not specific for evaluating lumbar disc herniations, is nevertheless highly sensitive to this disorder. The thermogram does not emit radiation to the patient and has no known negative side effects. As such, the thermogram may be useful in evaluating the need for further diagnostic imaging in the low back pain patient. The historical and clinical findings are essential in understanding the underlying reason for the patient's symptomatology.

References

1. Cailliet R: *Low Back Pain Syndrome*, ed 3. Philadelphia, F.A. Davis Company, 1981.
2. Bernard T, Kirkaldy-Willis WH: Recognizing specific characteristics of nonspecific low back pain. *Clin Orthop* 217:266–280, 1987.
3. Morris E, DiPaola M, Vallance R, Waddel G: Diagnosis and decision making in lumbar disc prolapse and nerve entrapment. *Spine* 11:436–439, 1986.
4. Wiesel S, Tsourmas N, Feffer H, Citrin C, Patronas N: A study of computer-assisted tomography: the incidence of positive CT scans in an asymptomatic group of patients. *Spine* 9:199–203, 1984.
5. Wexler C: *Atlas of Thermographic Lumbar Patterns*. Tarzana, CA, Thermographic Services, Inc, 1984, pp 9, 42–44.

12

Digital Thermography: Pre- and Post-Surgical Laminotomy and Diskectomy*

Harold W. Farris, DC, FCTS

HISTORY

A 39-year-old aircraft technician had low back pain in the past. These episodes, which were rare, resolved themselves spontaneously.

Two months ago, as he was helping lift a heavy object while helping someone move their household, he developed low back pain which later radiated into the right buttock, lateral thigh, calf, and foot.

The pain has increased with Valsalva maneuvers, causing difficulty in sleeping at night. There are associated tingling sensations, but no muscular weakness. The left leg is totally asymptomatic. The patient has no bladder or bowel problems and no neck problems.

Bedrest, pain medications, physical therapy, and chiropractic treatments had given momentary relief, but no lasting benefits.

PHYSICAL EXAMINATION

The range of motion of the back is limited. The patient carries himself with an overall torso list to the left. There are paraspinous muscle spasms and tender-

*Reprinted with permission from *The Thermal Compendium* 2(7):1–4, 1988.

ness to palpation at the lumbosacral junction, the right posterior iliac crest, and the sciatic nerve.

The patient's calves measure 36 cm. Straight leg raising is positive on the right at 10°, on the left 30°. There is crossed straight leg-raising sign. There is no common peroneal nerve tenderness.

MOTOR EXAMINATION

Strength is normal in hip flexors, quads, hamstrings, tibialis anterior, extensor hallucis longus, and peroneus and brevis. He can walk on his heels and toes.

SENSORY EXAMINATION

There is no sensory loss. Deep tendon reflexes are 2+ in knees and ankles.

REVIEW OF FILMS

Plane anterior–posterior and lateral spine radiograms show five lumbar vertebrae, with no spondylosis or spina bifida. An MRI scan shows a herniated disc, with free fragment at L5-S1, impinging on the right S1 nerve root.

THERMOGRAPHIC EVALUATION

Pre-surgical thermographic studies show focal heat emissions over L5 (Fig. 12.1) and hypothermia over the S1 dermatome (Fig. 12.2). The posterior lower leg thermogram (Fig. 12.2) shows a differential temperature between A

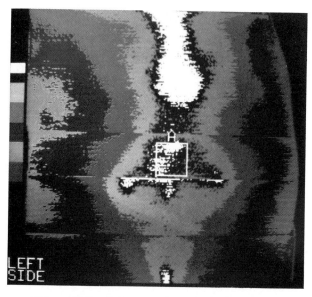

Figure 12.1. Pre-surgery lower back thermogram.

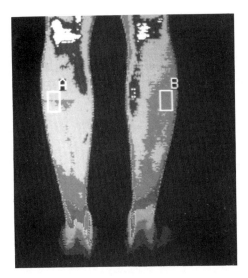

Figure 12.2. Pre-surgery posterior leg thermogram.

and B of .5°C; the dorsal feet thermogram (Fig. 12.3) shows a differential temperature between A and B of .8°C.

DIAGNOSIS

Right radiculopathy L5-S1. Patient was referred for diskectomy surgery.

POST-SURGICAL THERMOGRAPHIC EVALUATION

The posterior low back area demonstrated diminished bilateral L5 facet hyperthermia (Fig. 12.4). However, because of the surgical procedure, there

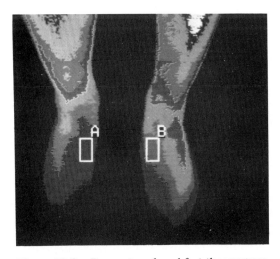

Figure 12.3. Pre-surgery dorsal feet thermogram.

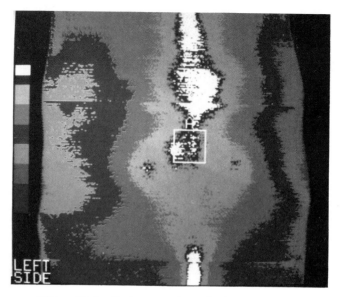

Figure 12.4. Post-surgery lower back thermogram.

were increased thermal emissions over the lumbar stripe, which was anticipated.

Multiple views of the extremities were performed and compared to objectify and quantify dermatomal regions and the effects of the surgery.

The plantar feet views, assessing the S1-S2 dermatome, remained at a differential temperature of .6°C as compared to the original study, showing no measurable change in dermatomal response. However, the L5 dermatome distribution over the plantar region of the foot did demonstrate improvement, the differential temperature decreasing from .9°C to .7°C, a .2°C improvement.

The dorsal feet studies of the L4-L5 dermatome demonstrated significant improvement—a pre-surgical differential temperature of .8°C (Fig. 12.3) improving to a post-surgical differential temperature of .1°C (Fig. 12.5).

The posterior lower leg thermograms showed a significant pre- and post-surgical difference. The pre-surgical thermogram (Fig. 12.2) showed a differential temperature of .5°C, the right being cooler. The post-surgical thermogram (Fig. 12.6) shows that the right calf is now .3°C *warmer* than the left, a change of .8°C from the pre-surgical thermogram.

CONCLUSIONS

This case confirms two of digital thermography's advantages: it is a directional test which affords timely documentation of the segmental level and extent of injuries and which aids the physician in prescribing the appropriate course of treatment; it is also an ideal objective and demonstrative tool in assessing the effectiveness of patient's response to treatment.

In this particular case of pre- and post-surgical evaluations, thermography

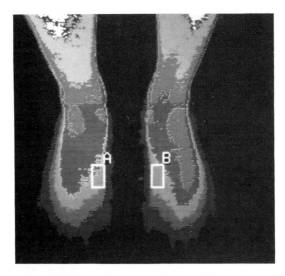

Figure 12.5. Post-surgery dorsal feet thermogram.

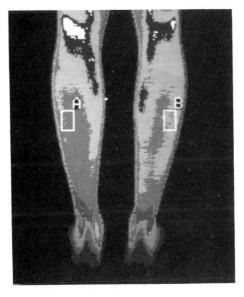

Figure 12.6. Post-surgery posterior leg thermogram.

gives the physician not only a visual interpretation of the physiological response to surgeries, but also a precise quantification that enables him to
more objectively determine treatment requirements and permanent residuals.

Digital thermography remains unique among all other diagnostic modalities in its ability to record and document, in a more visible manner, the
patient's neurophysiological responses to injury and treatment.

13

Thermography in Assessing the Effectiveness of Treatment*

Harold W. Farris, DC, FCTS

Abnormal thermograms occur in conjunction with vasomotor dysfunction. This dysfunction cannot be demonstrated by radiographic studies. Often it cannot be objectified by EMG or other conventional tests, particularly in nonsegmental and nonradicular pathologies, such as conditions involving muscles, tendon sheaths, ligaments, joint capsules, joints, skin, and subcutaneous tissue.

It is now widely recognized that mixed peripheral nerves contain separate, individual components of sensory fibers, motor fibers, and autonomic fibers, yielding myriad, varying results when tested by different clinical and electronic methods. Of these, thermography examines the sympathetic autonomic component. Consequently, a 100% correlation with the motor and sensory test of a dysfunctional mixed peripheral nerve should not be expected.

Thermography's role in early diagnosis—and particularly in monitoring throughout treatment and rehabilitation—is gaining increasing favor because successful therapy for muscular and fascial injuries depends on prompt, aggressive treatment in the acute stages. Delay in appropriate therapy leads to the development of additional problems, such as myofascial pain syndrome, chronic causalgia, and chronic sympathetic dysfunction, decreasing the chances for complete healing or prolonged recovery.

The case presented is that of a 43-year-old man involved in a work-related

*Reprinted with permission from *The Thermal Compendium* 2(5):25, 1988.

injury on October 10, 1978, which resulted in disc rupture at L5-S1. The patient underwent spinal surgery with some degree of success; however, he still complains of headaches and neck pain, low back pain, pins and needles in his arms and legs, and the feeling of numbness in the left hand. The patient was evaluated at the request of his treating physician before receiving epidural block and manipulative procedures in relation to the compensatory cervicothoracic myofascial syndrome.

A series of pre-treatment thermograms were made on June 9, 1988. The lumbar thermogram (Fig. 13.1) demonstrated focal hyperthermia midline at L5-S1, with a general decrease of heat emissions over the right buttock. The posterior cervicothoracic thermogram (Fig. 13.2) showed increased heat emissions extending over the paravertebral musculature in the cervicothoracic region, extending into the left superior scapular border on the left. The differential temperature between areas A and B is .2°C. The ventral hand thermogram (Fig. 13.3) demonstrated a general decrease of thermal emissions over the left hand, with a mean temperature difference of 1.7°C between the fifth digits, and .8°C between the fourth digits.

Initial thermograms represent motor unit dysfunction, posterior primary nerve root irritation at L5-S1, and subtle compression of the ventral nerve root, which is demonstrated by the colder right buttock. The cervicothoracic area represented osteoligamentous irritation with secondary neurovascular reflexia or dysfunction of the left upper extremity.

Following treatment procedures, the patient returned on June 14, 1988. Post-treatment thermographic evaluation was obtained. Lumbar thermograms (Fig. 13.4) demonstrated some reduction in thermal emissions over the L4-L5 anatomical level. However, the L5-SI motor unit appeared unchanged. The buttocks area did demonstrate some improvement in thermal

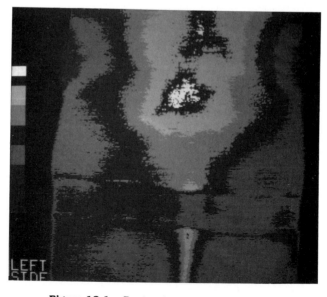

Figure 13.1. Pre-treatment posterior lumbar.

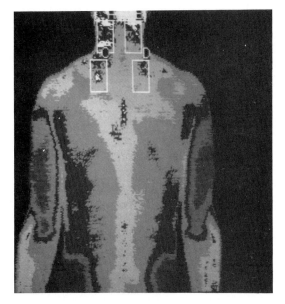

Figure 13.2. Pre-treatment cervicothoracic.

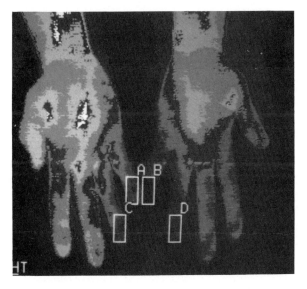

Figure 13.3. Pre-treatment ventral hands.

Figure 13.4. Post-treatment posterior lumbar.

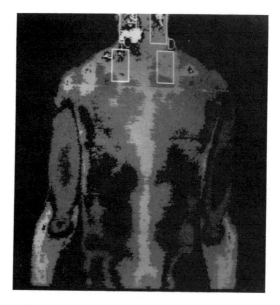

Figure 13.5. Post-treatment cervicothoracic.

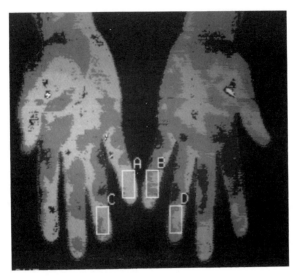

Figure 13.6. Post-treatment ventral hands.

symmetry. Of interest, the patient stated that his subjective complaints of the low back and legs seemed unchanged.

Posterior cervicothoracic thermograms (Fig. 13.5) demonstrated improvement in the cervical region, returning thermal symmetry. (The differential temperature between areas A and B is now 0°C.) However, the myofascial patterns remained. Of greater consequence was the post-treatment ventral hand studies (Fig. 13.6), which demonstrated an improvement of the differential temperatures between the fifth digits, now at .3°C, and between the fourth digits, now at .1°C.

Diagnosing and monitoring neurovascular and musculoskeletal abnormalities by thermography is based on thermal asymmetry between normal and abnormal sites. It is well suited for the physician's decision-making process, because it is a highly sensitive, reliable, and objective means of early diagnosis, prognosis, and management of such conditions.

14

Digital Thermography in the Assessment and Treatment of Hypersympathetic-Maintained Pain Syndrome*

Harold W. Farris, DC, FCTS

HISTORY

Four months after an automobile accident in which she received injuries to the cervical spine and left shoulder, a 34-year-old woman presented herself for examination and treatment. The patient is left-handed.

At the time of examination, the patient had been and continued to be disabled from her work because of the severity of her pain complex.

Previous treatment consisted, for a six-week period, of daily physical therapy, which included moist heat, massage, and biofeedback. Treatment requirements were based on an initial diagnosis of strain or sprain of the cervical spine and cervical or shoulder syndrome secondary to the motor vehicle accident as reported.

---

* Reprinted with permission from *The Thermal Compendium* 2(4):1988.

CLINICAL FINDINGS

The following tables present the pertinent clinical findings:

Motion (Cervical)

	Norm	ROM	Cause
Flexion	45	35	Pain, spasm
Extension	55	30	Pain
Left Rotation	70	40	Pain, stiffness
Right Rotation	70	45	Pain, stiffness
Left Lateral Flex	40	35	Stiffness
Right Lateral Flex	40	30	Stiffness

Neurologicals

Sensation testing (pain and touch)

Root Level	Left	Right
C5	2	—
C6	2	—

Orthopedics (Cervical)

Cervical Compress	Bilateral
Cervical Distract	Bilateral
Adsons	Negative
Costoclavicular	Negative
Shoulder Depress	Left
Valsalva	Positive
Soto Hall	Negative

Dynamometer

	Left	Right
First Try	20	35
Second Try	25	45
Third Try	25	55

Deep Tendon Reflexes

	Left	Right
Biceps	Normal	Normal
Triceps	Normal	Normal
Brachioradialis	Normal	Normal

[All reflexes were brisk.]

RADIOLOGIC STUDIES

Multiple views of the cervical spine revealed no conclusive radiographic evidence of recent gross pathology, luxation, or fracture. Bone quality and density was essentially normal for a 34-year-old woman. The cervical lordotic curve appeared normal. Cervical flexion and extension studies did demonstrate alteration on George's line at C5 and C6.

THERMOGRAPHIC EVALUATION

On March 9, 1988, thermography studies of the upper extremities demonstrated a general hypothermia of the left arm, forearm, and hand as compared to the right. The hands study (Fig. 14.1) showed differential temperatures of 1.5°C between areas A and B and .3°C between areas C and D. Thermographic evaluation was compatible with a hypersympathetic maintained pain syndrome of the left upper extremity, secondary to the motor vehicle accident as reported.

DIAGNOSIS AND MANAGEMENT

Post-traumatic hypersympathetic maintained pain, left upper extremity. Treatment by electro-muscle stimulation, deep myofascial massage, and cervical distraction.

POST-THERMOGRAPHIC EVALUATION

On May 25, 1988, a repetition of thermography studies of the upper extremities demonstrated a favorable treatment response. The differential temperatures in Figure 14.2 are .3°C between areas A and B and .2°C between areas C and D.

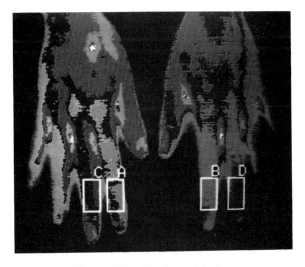

Figure 14.1   Hands on 03/09/88.

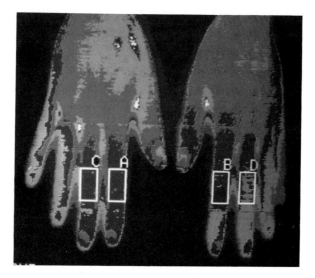

Figure 14.2 Hands on 05/25/88.

CONCLUSIONS

This case provides a good illustration of how the symptomatology of certain conditions can evade most diagnostic modalities. Thermography remains the most sensitive and accurate test for the early detection of sympathetic dysfunction—especially hypersympathetic maintained pain syndrome, and it most objectively quantifies treatment effectiveness during case monitoring of ongoing treatment.

15

Thermography and Sensory Neuropathy*

Harold W. Farris, DC, FCTS

HISTORY

A 41-year-old male executive complained of constant neck pain, coldness, and numbness of his hands and fingers, and numbness of his left ankle and foot. His symptoms began after a fall while ice skating six months before his first examination.

Previous treatment consisted of chiropractic adjustments and physical therapy, including cervical traction 10–15 minutes daily. His condition continued to worsen, and he was examined by his medical physician, who prescribed diazepam.

CLINICAL FINDINGS

Radiographic studies: A neutral lateral cervical view (Fig. 15.1) demonstrated a decreased vertebral body height at C4, with questionable trophic findings at C5-C6.

Physical examination demonstrated a significant loss of cervical motion in all planes, the left and right lateral flexion being most pronounced.

Dynamometer testing showed a 10 to 15 point difference, the right being the weakest. (The patient is right-handed.) Deep tendon reflexes were essentially normal.

Sensation testing for pain and touch was positive in the C5-C6 dermatomes.

---

*Reprinted with permission from *The Thermal Compendium* 2(3): 1988.

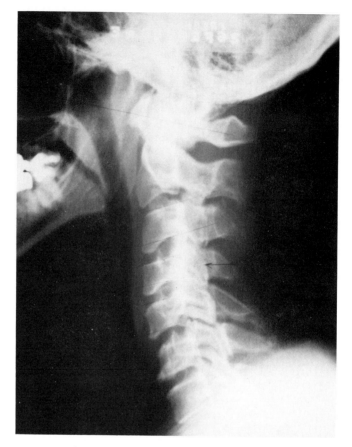

Figure 15.1. Neutral lateral cervical view.

Palpation produced pain at the suboccipital upper and lower cervical and upper thoracic paravertebrals.

Magnetic resonance imaging (Figure 15.2): The radiologist reported that "T1 weighted sagittal and gradient refocussed axial views suggest some stenotic change beginning at the C3-C4 level and extending to the C4-C5 region. This is felt to also include some posterior encroachment by bony and ligamentous changes. No definite features of a disk herniation are identified at this time." The radiologist's report concluded there were "Some stenotic changes beginning at the lower portion of C3 and extending to at least the upper portion of C5."

THERMOGRAPHIC EVALUATION

Thermographic evaluation demonstrated thermal asymmetry over the cervical spine, producing a general blush and increased heat emissions with a

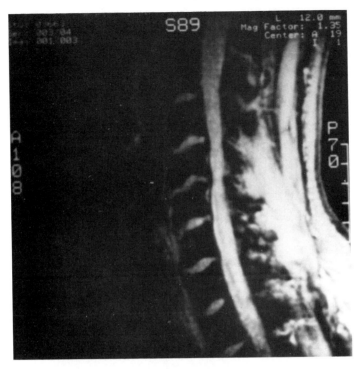

Figure 15.2. Magnetic resonance imaging.

circular, well-defined area of decreased heat emissions over the midline of the midcervical components.

A thermograph of the thoracic region demonstrated a general increase of heat emissions extending over the right paravertebral musculature (Fig. 15.3).

The dorsal hand studies (Fig. 15.4) demonstrated thermal asymmetry, with a differential temperature of 1.4°C, the right hand being the coldest.

The posterior and anterior leg studies (Figs. 15.5) demonstrated a marked stocking appearance, extending over the left ankle and foot, with a differential temperature of as much as 3.9°C.

DIAGNOSIS AND MANAGEMENT

Sensory neuropathy, idiopathic etiology, C3-C5.

The patient will continue conservative therapy to reduce secondary myofascial involvement in the cervicothoracic region and will be referred for further neurological evaluation, e.g., lumbar myelogram, or nerve fiber biopsy, or both. Prognosis remains guarded.

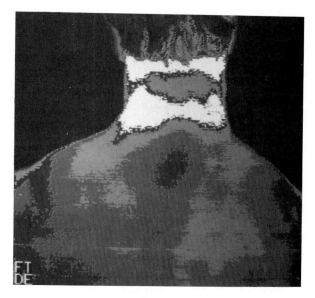

Figure 15.3. Increased thermal emissions over the right thoracic paravertebral muscles.

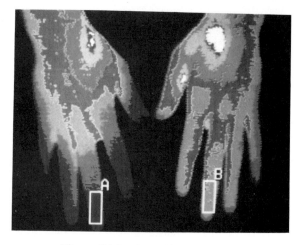

Figure 15.4. Dorsal hand studies.

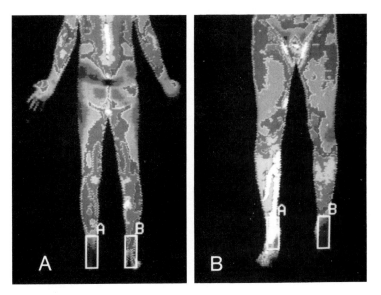

Figure 15.5. **A,** Posterior leg study. **B,** Anterior leg study.

CONCLUSION

This case demonstrates the value of thermography in the clinical setting. Thermography manifested a definite sensory autonomic dysfunction. Without this information, the patient would probably have continued on a regimen of incomplete, misdirected, or unsuccessful treatment. As a directional test, thermography quantified the need for further diagnostic interpretation.

16

Clinical Sensitivity and Thermographic Accuracy*

Harold W. Farris, DC, FCTS

Proper functioning of a vertebral motor unit requires synergy of the involved muscles. An awkward or unexpected movement may lead to an improper division of forces. Some elements of a motor unit may then undergo traction and compression greater than the restrictive capacity (1), resulting in stretching or tearing, or both, of any or all of the soft tissue support structures, stimulating nociceptive pain fibers, the sinuvertebral nerve fibers, and ventral and autonomic nerve fibers. Such stimulation results in a visually observable cutaneous vasomotor response which can be objectively quantified by digital thermography, thus providing the physician with a diagnostic potential for evaluating primarily motor unit complexes and their physiological components.

The concept of the intervertebral motor unit was initially described in Schmorl and Junghanns (1). Further investigation by Wall and Melzack (2) classified three types of dysfunction in the motor unit:

- Involvement of the entire motor segment
- Involvement of the posterior aspect of the motor unit
- Involvement of the foraminal confines.

Clinical studies using thermography have correlated and established a high degree of accuracy and sensitivity in determining the physiological aspects and locations of primary intervertebral motor unit irritations or dysfunctions.

---

*Reprinted with permission from *The Thermal Compendium* 2(6): 1988.

HISTORY

A 22-year-old woman employed as a stock person injured her back while lifting boxes of clothing weighing from 60 to 80 pounds. She experienced sharp low back pain that radiated into the left hip. The patient initially sought chiropractic treatment; however, because of increasing pain in the low back and left hip, she was admitted to an emergency medical clinic, which prescribed physical therapy three times per week for eight weeks and medication to relieve myospasms and reduce pain.

At the end of the physical therapy, her complaints persisted, and she began experiencing radicular complaints into the left lower extremity. She was referred to a second chiropractor for further evaluation and treatment. He diagnosed a low back L5 strain and treated it conservatively with spinal manipulations and ultrasound, with only a minimal degree of success.

TOMOGRAPHIC STUDY

After receiving treatment for about six months, she was referred for orthopaedic evaluation, which proved to be negative. She was subsequently referred for neurological and further diagnostic evaluation. A computed tomographic scan of the lumbar spine was reviewed by two different radiologists on separate occasions.

The radiologist at the imaging laboratory observed that the images revealed a herniated nucleus pulposus laterally at L5-S1. Two weeks later, a second radiologist reviewed the same images and found that "the degree of bulging [at L5-S1] does not appear to be significant on the scan" and that the study was "negative."

Because of these mixed impressions, electrodiagnostic studies were performed; they were interpreted as normal for L5-S1 radiculopathy. However, the patient's intolerance to the testing prevented completion of the studies. At this time, the original examining doctor suggested that the patient was malingering and that she be referred for psychological counseling.

At the persuasion of the treating chiropractor, thermographic evaluation was performed.

THERMOGRAPHIC EVALUATION

May 18, 1988: The posterior low back thermogram (Fig. 16.1) showed focal hyperthermia approximating the left L5 facet and a midline focal hyperthermia approximating anatomical level S1-S2.

Lower extremity studies appeared symmetrical except for subtle hypothermia over the dorsum of the left ankle as compared to the right. However, the plantar feet studies (Fig. 16.2) showed a significant thermal asymmetry of the left heel as compared to the right, with a temperature difference of 4.6°C, the left being hypothermic.

These studies indicated a cutaneous distribution of the tibial nerve S1-S2, demonstrating dermatomal vasomotor reflex. Initial thermographic evalua-

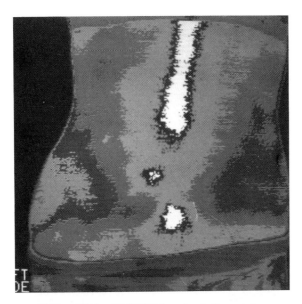

Figure 16.1. 05/18/88 Low back thermogram.

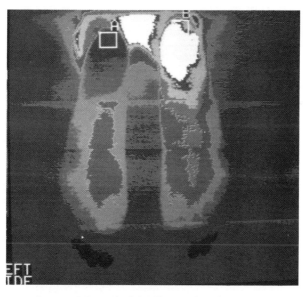

Figure 16.2. 05/18/88 Plantar feet thermogram.

tion documented L5 facet syndrome and S1-S2 motor unit dysfunction with radiculopathy.

Physical examination revealed a pelvic imbalance of approximately one inch, with resultant contractured left leg of approximately ¾ inch.

Because of these findings, the thermologist recommended additional neurological evaluation to assess S1-S2 radiculopathy, and further chiropractic and podiatric evaluation to assess the possible effectiveness of an orthotic support to gain mechanical balance of the lumbar spine.

The second neurological evaluation confirmed the presence of a tibial nerve S1-S2 radiculopathy.

Podiatric treatment consisted of application of a ½-inch orthotic support, which improved the mechanical balance of the lumbar spine.

July 26, 1988: Further thermographic evaluation, to objectify the pa-

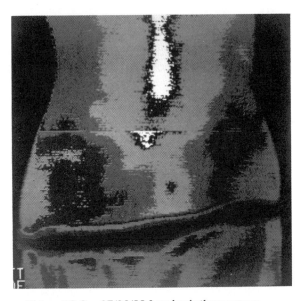

Figure 16.3. 07/26/88 Low back thermogram.

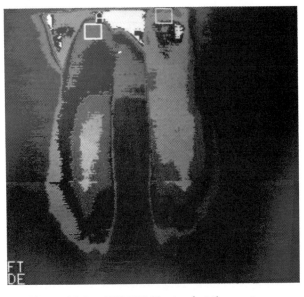

Figure 16.4. 07/26/88 Plantar feet thermogram.

tient's response to treatment revealed a marked improvement of the thermal characteristics of the S1-S2 motor unit (Fig. 16.3). The plantar feet studies (Fig. 16.4) showed a significant improvement in the vasomotor dysfunction secondary to tibial nerve radiculopathy, with a temperature differential reduced to 2.7°C.

The patient is now continuing conservative chiropractic treatment and showing progressive and steady improvement in her symptomatology.

CONCLUSION

This case proves the value of thermography in diagnosing and managing motor unit dysfunction syndromes. Because of thermography's demonstrative characteristics, it can quickly and easily locate the type of motor unit irritation or injury. Thermography serves as an ideal screening procedure in patients being considered for additional diagnostic procedures.

References

1. Schmorl G, Junghanns H: *The Human Spine in Health and Disease.* New York, Grune and Stratton, 1971.
2. Wahl P, Melzack R: *Textbook of Pain.* New York, Churchill Livingstone, 1984.

17

Thermography in Assessing Injury Risk Factors*

Harold W. Farris, DC, FCTS

HISTORY

A 27-year-old construction worker with an acute lower back injury was referred for evaluation and treatment by his employer. He had a history of three injuries to his lower back within the preceding 18 months. His initial injury resulted from lifting a troweling machine while pouring a house foundation. He received chiropractic treatment and physical therapy and returned to work in one week.

His second low back injury occurred approximately five months later, while he was attempting to lift concrete blocks. This injury resulted in sharp, deep pain over the L4-L5 facets with bilateral projection of this pain into the buttocks. He received chiropractic treatment and physical therapy for a lumbar strain syndrome and returned to work in four weeks. He expressed residual complaints for three to four months.

His third injury occurred nine months later, while he was attempting to pick up a sack of cement. This injury resulted in debilitating pain of the low back with radicular pain extending into the right leg, calf, and foot.

CLINICAL FINDINGS

Injury 1/12/87. Physical examination revealed a slender, well-developed muscular male who appeared in mild to moderate distress.

---

*Reprinted with permission from *The Thermal Compendium* 2(2): 1988.

Positive findings included decreased lateral flexion of the lumbar spine with difficulty in returning to an upright stance after guarded forward bending. Neurological tests and radiographic studies were negative.

Injury 6/11/87. Neurological examination revealed mild hypesthesia over the L4 dermatomes on the right. Reflexes were normal and the remainder of the physical examination was normal.

Injury 3/2/88. Neurological examination revealed hypesthesia over L2, L3, L4, L5, and S1 dermatomes on the right and absence of the Achilles reflex.

THERMOGRAPHIC FINDINGS

Injury 1/12/87. Multiple studies of thermograms were performed of the back and lower extremity following 20 minutes of equilibration in an ambient temperature of 20°C (Figs 17.1 and 17.2). Figure 17.1 shows mean temperatures of 31.4°C on the left and 30.4°C on the right, a differential of 1.0°C.

The posterior lumbar studies demonstrated focal hyperthermia over the L4-L5 facets on the left. The posterior leg studies were symmetrical. The dorsal foot studies demonstrated hypothermia over the L4 dermatome on the right.

Thermographic studies were suggestive of L4-L5 facet syndrome with hypersympathetic (sensory) nerve distribution L4, right.

Injury 6/11/87. Thermographic evaluation was indicated, but not authorized.

Injury 3/2/88. Same protocol as first thermographic examination.

The posterior lumbar studies demonstrated multiple motor unit involvement (Fig. 17.3).

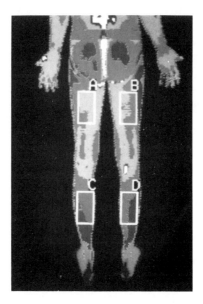

Figure 17.1. Lower extremity thermogram.

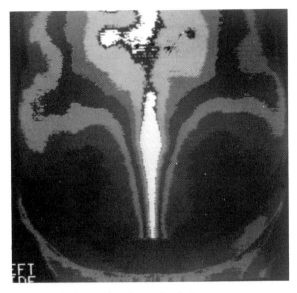

Figure 17.2. Lower back thermogram.

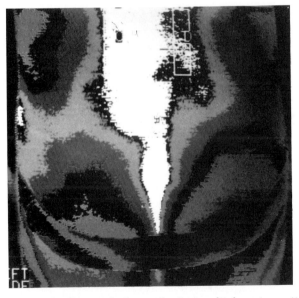

Figure 17.3. Posterior lumbar study demonstrating multiple motor unit involvement.

The posterior leg studies demonstrated definite thermal asymmetry. Figure 17.4 shows mean temperatures of 35.9°C on the left and 34.6°C on the right, a differential of 1.3°C.

Thermographic interpretation suggested a hypersympathetic right (sensory) nerve distribution at L4, L5, and S1, with marked involvement of the lower roots.

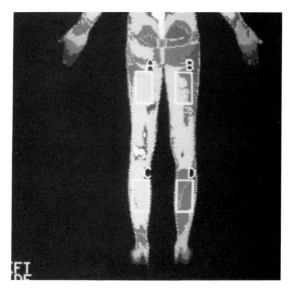

Figure 17.4. Note the thermal asymmetry.

MAGNETIC RESONANCE IMAGING

Two sequences, consisting of sagittal and axial scans, showed reduced signal intensity from the L4-L5 and L5-S1 disks consistent with desiccation. Material of relatively low signal intensity projects to the right at L5-S1, obviously distorting the nerve root at that point (arrows, Figure 17.5). No abnormal areas of high signal intensity and no mass lesions were identified.

The radiologist's impressions were "Degenerative disk disease at L4-L5

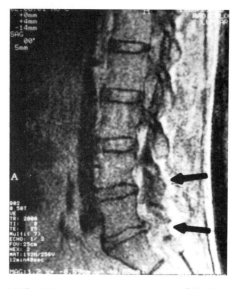

Figure 17.5. Magnetic resonance image of the involved area.

and L5-S1" and "evidence of a nucleosis proposis hernia on the right at L5-S1."

DIAGNOSIS AND MANAGEMENT

Central disc involvement, L5, S1. The patient is scheduled for neurosurgical evaluation.

CONCLUSION

Thermography is a reliable, noninvasive, low-cost physiological test procedure highly suited for pre-employment screening and initial injury evaluation. The thermograms and their interpretation become a permanent part of the employee's medical file.

This case demonstrates the value of using thermography to establish, document, and assign a "risk factor" to a patient's future activity based upon the thermal findings, when routine evaluations were normal. It has become clear that thermography as a physiological test often offers the only source of objective data for quantifying and qualifying injuries or risk factors of potential exacerbation.

18

Low Back Pain and Radiculopathy Confirmed by Thermography and Magnetic Resonance Imaging in the Absence of Significant Physical Findings

Kim R. Hoover, DC

HISTORY

The patient, a 49-year-old black male truck driver and heavy equipment operator, presented on September 2, 1988, complaining of left-sided neck pain and limitation of neck movement, radiating pain over the left upper back (shoulder) into the left hand, and dull, constant low back pain which became severe when sitting. The patient's pain had resulted from a glancing, head-on automobile collision in which he was thrown from side to side. He was initially aware of the pain and injury to his left arm, hand, shoulder, and face. His face, shoulder, and left thumb were swollen. He received stitches in his left thumb and left arm, and his neck and back were x-rayed during his emergency visit to the hospital. The emergency room physician referred the patient to a local orthopaedist, who prescribed cyclobenzaprine hydrochloride. The patient did not return to the orthopaedist for follow-up care. Neither did he return to work during the month between his automobile injury (August 2, 1988) and his initial chiropractic visit.

CLINICAL FINDINGS

The patient initially showed no antalgic posture, gait abnormality, or difficulty with change of posture. He exhibited minimal limitation of left lateral lumbar flexion, and minimal pain with passive and active range of motion to all ranges. Heel and toe walk were performed without difficulty. Kemp's test was positive on the left side. The sitting straight leg raise was unremarkable; Lasaque and Braggard's tests were negative. The patient performed the supine leg raise and lower. Lindner's test produced pain in the lumbar area. His reflexes were symmetrical and brisk; Mennell's test was negative, as was hip extension and Nachlas and Eli tests. The patient did have grade 4 weakness of his abdominal muscles. All other manual muscle testing was unremarkable. Valsalva was negative. Although the patient exhibited lumbar fixation, perhaps his most significant clinical finding was bilateral pain over the fourth and fifth lumbar levels with paraspinal muscle spasm. The patient reported improvement: his lower back pain had lessened in intensity and his upper extremity symptoms had resolved. He was referred for evaluation and rehabilitation and showed immediate improvement in his strength and range of motion. He returned to work on January 9, 1989. One week later, he returned to the office in more severe low back pain, experiencing some weakness into his right knee. The patient showed little change in his clinical picture from the initial evaluation. Right-sided muscle spasm was evident. He continued to find sitting for more than 10 minutes painful.

THERMOGRAPHY

Serial thermograms of the low back and lower extremities were exposed on October 19, 1988 (Fig. 18.1). Follow-up thermographic study was performed on November 18, 1988, consisting of lumbo-pelvic views only (Fig. 18.2). A complete study was repeated on January 20, 1989. A pronounced vertical stripe of increased heat emission could be seen extending from the right lower lumbar region. This lumbar stripe is often consistent with a ruptured lumbar disc (1–9). Thermographic study also suggested left S-1 nerve fiber irritation.

RADIOLOGICAL STUDIES

Lumbo-sacral anteroposterior and lateral x-rays showed a right pelvic deficiency and compensating lumbar scoliosis. The patient had a flexion subluxation at the third lumbar with degenerative changes visualized at the third and fourth lumbar disk (L3-4). Magnetic resonance imaging (MRI) was ordered in response to the patient's persistent subjective and objective find-

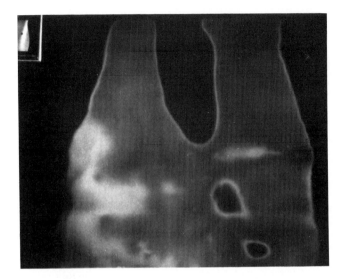

Figure 18.1. Thermogram study of 10/19/88.

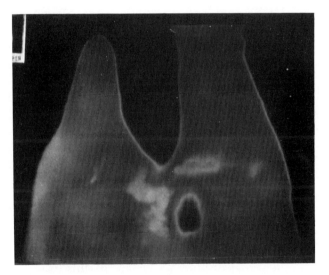

Figure 18.2. Thermogram study of 11/18/88.

ings. The L3-4 disc was narrowed and a broad-based posterior herniation of the disc was evident (Fig. 18.3). The herniation was causing an anterior extradural defect compressing the thecal sac. According to the radiologist, the "protruded disc material extends into the inferior portions of the bilateral neuroforamina, and there may be some distortion of the right L-3 nerve root as it exits the foramina."

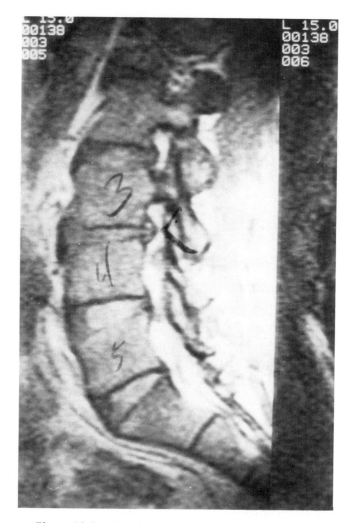

Figure 18.3. MRI showing evidence of disc involvement.

COMMENT

In most cases, a diagnosis of lumbar disc herniation may be made on the basis of clinical findings alone. In this case, however, more weight should have been given to the thermographic findings. MRI was not ordered until after the January study was completed.

References

1. Lerman VJ: Personal communication to the author, 1983.
2. Wexler CE: Lumbar, thoracic, and cervical thermography. *J Neurol Orthoped Surg* 1:37–41, 1979.
3. Ching C, Wexler CE: Peripheral thermographic manifestations of lumbar-disk disease. *Appl Radiol* 100:53–58, 1978.
4. Tichauer B: Objective corroboration of back pain through thermography. *J Occup Med* 19:727–731, 1977.

5. Kamajian II GK, Tilley P: Thermography of the back in asymptomatic subjects. *JAOA* 74:429–431, 1975.
6. Green J, Noran WH, Coyle MC, Gildemeister RG: Infrared Electronic thermography (ET): a noninvasive diagnostic neuroimaging tool. *Contemp Orthop* 11:29–36, 1985.
7. Harpman HL, Knebel A, Cooper CJ: Clinical Studies in thermography. *Arch Environ Health* 20:92–100, 1970.
8. Heintz ER, Goldberg HI, Taveras JM: Experiences with thermography in neurologic patients. *Ann NY Acad Sci* 121:177–189, 1964.
9. Dudley WN: Distal thermographic residuals: double lesion neuropathy (submitted for publication).

19

Thermographic Findings in a Patient with Lymphoma of the Upper Lumbar Spine

Geoffrey Gerow, DC, DABCO, Michael Poierier, DC, and James Christiansen, PhD

When evaluating a patient with a traumatic medical history, it is often easy to attribute symptoms to mechanical dysfunction. Occasionally a patient presents with symptoms that may on the surface appear mechanical, but on closer examination will reveal an underlying pathology. We describe a case of a patient with upper lumbar lymphoma who received a thermogram during his diagnostic evaluation.

Lymphoma is a malignancy of the lymphatic system. Although lymphatics are found throughout the body, a predilection for this disease exists in the lymph nodes, gut, lungs, and skin. The typical patient with lymphoma presents with a painless neck mass, the result of an involved cervical lymph node. Additional expected findings are drenching night sweats, remittent fever, weight loss, and splenomegaly. A patient may present without obvious lymphatic involvement but with spinal pain secondary to a bony lesion. Trauma to the involved area may "expose" the disease. Lymphoma of the bone is relatively uncommon (1).

The mean overall age at which lymphoma of the bone occurs is 46.2 years. The male to female ratio of occurrence is 1.6 to 1. Although the classic radiographic spinal manifestation of lymphoma is the sclerotic "ivory vertebra," the more common lesion is osteolytic. In advanced disease, spread to the spinal canal may cause spinal cord compression. The extent of involvement is determined by physical examination of all accessible lymphatics, renal and liver function studies, imaging of the chest and abdomen, bone and liver-spleen radionuclide scans, and bilateral lower extremity lymphography. Definitive diagnosis is by biopsy. Staging of lymphoma is based on the location of involved sites and whether the patient has systemic symptoms. The 5- and 10-year survival rates in a sample size of 422 patients from 1907 to 1982 were 58% and 53%, respectively. Within that group, the 5- and 10-year survival rates of patients with bone and nodal or soft tissue disease were 22% and 12.5%, respectively.

CASE STUDY

The patient, a 46-year-old white male, presented to a chiropractic college clinic 9 days after he had moved a large number of picnic tables. His physician prescribed acetaminophen (300 mg) with codeine phosphate (30 mg) every 3 hours and cyclobenzaprine hydrochloride three times a day, which for the week following his visit provided the patient little to no relief. He had not been hospitalized previously and was a nonsmoker. The patient described his pain as being in the low back area, radiating into the posterior aspects of both thighs to the level of his knees. He experienced the greatest pain when flexing forward to put on his socks.

A standard physical examination and thermographic screen of the patient's back during his initial visit revealed hyperthermia about the lower and upper lumbar spinal regions (Fig. 19.1).

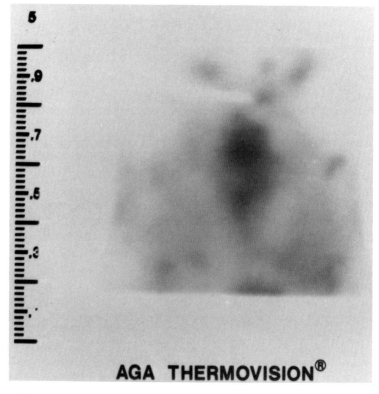

Figure 19.1. Thermogram showing hyperthermia in lower and upper spinal regions.

A lumbosacral x-ray series revealed a 15% spondylolisthesis of the fifth lumbar segment and a compression fracture of the first lumbar (L1) segment (Fig. 19.2). Linear tomograms of the thoracolumbar region confirmed this evaluation.

A complete blood count and urinalysis performed on admittance yielded values within normal limits. A blood chemistry was taken the same day, revealing elevations in AST, ALT, LDH, GGTP, ALK phos, and total globulin. Protein electrophoresis showed high alpha$_1$ and alpha $_2$ and beta globulin. No M-spike was present. Fecal material was negative for occult blood. The patient was referred for further work-up.

Eight days following the initial presentation to the clinic, a radionuclide bone scan was performed, revealing increased uptake of radionuclide in the upper lumbar spine in the projection of the L1 and L2 neural arches. This uptake may represent facet joint areas of increased uptake. The possibility of neoplasm is considered to be less likely, however, radionuclide imaging is not specific in morphologic diagnosis.

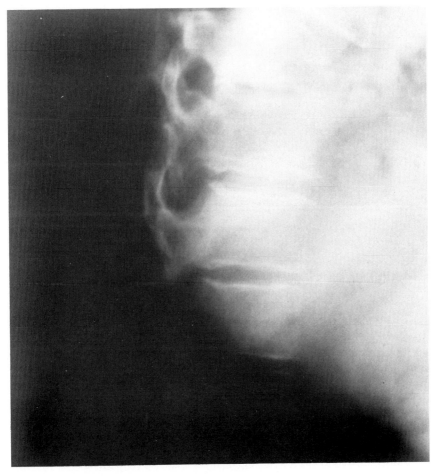

Figure 19.2. X-ray showing lumbar compression fracture.

The patient continued outside the clinic for almost 3 weeks, whereupon he returned complaining of anterior thigh weakness and paresthesia.

An orthopaedic consultation approximately 1 month following the thermographic evaluation revealed the lower extremity peripheral pulses and deep-tendon reflexes were within normal limits. Straight leg raise on the right side at 70° aggravated the patient's chief complaint and was unremarkable on the right. Palpation of the posterior musculature of the back and lower extremities revealed no tender areas. The patient demonstrated a minor's sign modification in going from a seated to a standing position. Ranges of motion of the thoracolumbar spine revealed flexion 15°, extension 10°, right lateral flexion 25°, and left lateral flexion 25°. All these motions were inhibited by pain in the back region. Thoracolumbar rotation was within normal limits bilaterally. The patient was able to walk with increased pain on his toes but pain prevented him from walking on his heels. Percussion of the spinous processes from C7 through L5 revealed no tenderness. Compression over the T12-L1 levels were extremely tender. Skin redness is present about the region of the L1 segment. Such redness was not visible when the thermogram was performed.) No motor loss was perceived.

The liver was palpated one-finger breadth below the right costal margin. Prostate examination revealed a small nodule within the left lobe. The patient indicated he has no sexual difficulties and no trouble starting or stopping his urine. A male Pap test was performed and was within normal limits. The patient demonstrated an erythrocyte sedimentation rate of 38mm/hour corrected.

Plain films were again taken 2 days after the orthopaedic consultation; the pictures showed progression of the fracture of L1 with now present involvement of T12. The patient was referred. Exploratory surgery revealed that the patient had a lumbar lymphoma in the anterior portion of L1, with metastasis to the liver. The patient has since died.

DISCUSSION

Although the thermographic findings did not point to diagnosis, they did lead to the suspicion that the T12-L1 region was one of increased heat. This area became visually hyperemic to the naked eye. The case is interesting in that the reason for the compression fracture at the L1 segment was not immediately apparent. The patient had had no symptoms associated with that level on the initial visit and complained only of low back pain and bilateral gluteal radiation. To complicate matters, he had had a history conducive to a diagnosis of lumbosacral strain or sprain, aggravation of the L5 spondylolisthesis, or possible lower lumbar discopathy. The lumbosacral x-ray series brought this disorder to the attention of the clinician. Had it not been on the film (e.g., had the lesion been higher in the spine), it might have taken much longer to isolate the area for investigation. In such a hypothetical case, on the basis of the thermographic and clinical findings, perhaps the clinician would have studied the region in question further. The chemistry evaluation led to the suspicion of liver and bone involvement. The protein electrophoresis reinforced this concern while making multiple myeloma a less likely diagnosis. The bone scan confused the issue. Because of the increased uptake of the technetium, however, the report favored a facet arthrosis. It was not until subsequent plain film radiography was performed that the spread of the lesion was inferred.

Reference

1. Ostrowski M, Unni K, Banks P, Shives T, Evans R, O'Connell M, Taylor W: Malignant lymphoma of bone. *Cancer* 58:2646–2655, 1986.

20

Three Possible Cases of Nonorganic Complaints

Geoffrey Gerow, DC, DABCO, Robert Shiel, PhD, and James Christiansen, PhD

Patients presenting with physical symptoms not based on substantive somatic disorder may demonstrate a nonorganic disorder. There are two categories of such disorders: malingering and psychogenic or psychological. Malingering is defined as simulating symptoms of an illness with the intent to deceive (1). The *Diagnostic Standards Manual III, Revised* (DSM-III-R) describes the essential feature of malingering as "intentional production of false or grossly exaggerated physical or psychological symptoms motivated by external incentives . . . " DSM-III-R further states that malingering should be strongly suspected if any combination of the following is noted: 1) the medicolegal context of presentation; e.g., the person has been referred by his or her attorney to the physician for examination; 2) a marked discrepancy exists between the person's claimed stress or disability and no objective findings; 3) a lack of cooperation during the diagnostic evaluation and in complying with the prescribed treatment regimen; and 4) the presence of antisocial personality disorder.

The detection of simulated complaints is anything but easy. An essential feature is false or exaggerated physical or psychological symptoms (1). Sometimes when speaking about the affected physical area, a patient will cease to move it; however, when discussing something else, the patient may move the area quite freely. It may be useful for the doctor to take the patient history to help in detecting such patients through redirection of questions, especially since patients who have been examined before may have prepared answers. Patients' use of terms to describe their conditions (particularly such terms as ripping, tearing apart, and exploding) are not commonly used when physical pain is present (1). In a study by Leavitt (2), subjects with and without low back pain were told to exaggerate their pain sufficiently to convince the doctor to take them off work. The subjects used 45 pain words to classify low back pain. Leavitt found that 83.6% of the volunteers without pain were simulating it. True malingering is found in patients with low back pain, but the incidence is relatively small, occurring in no more than 5% of cases (3) and is not present to any great degree in the work force (4). Exaggeration of symptoms may also be evaluated with symptom validity testing (5).

Psychogenic or psychological disorders are divided into several types: *conversion disorder*, which is characterized by a loss or alteration in function with a temporal relationship between symptoms and the environment; *somatoform pain*, which is pain of an extended duration; *somatization*, which is characterized by multiple symptoms without cause; and *undifferentiated somatoform*, which is characterized by physical complaints such as fatigue, loss of appetite, and gastrointestinal or urinary complaints of greater than 6 months, duration without supportive organic signs to suggest the symptomatic degree of dysfunction.

Attempts have been made to evaluate persons with factitious or psychogenic disorders. Seven basic concepts have been arrived at to make that determination. La Belle indifference is a term describing a patient's inappropriate indifference to the seriousness of his or her complaints (6). Although often found in patients with somatoform disorders, it is also seen in patients with psychoses, particularly schizophrenia and depression. Other findings characteristic of somatoform disorders include sensory loss that is not anatomically consistent, changing patterns of sensory loss on repeat examinations, change in sensory or motor findings at the doctor's suggestion, hemianesthesia splitting the midline, and a giveaway weakness on motor testing, as well as a unilateral loss of vibratory sense which is present even when two sides of the forehead or sternum are sequentially tested (6). The presence of these findings alone is not indisputable proof of somatoform disorder. A misdiagnosis of hysteria has been related reportedly to the following: person's sex, notably female prior psychiatric illness, psychodynamic possibilities which would explain the physical manifestations, and exaggerated physical findings (7). Hysterical paraplegia is associated with patients who are predominantly paraplegic, with histories of psychiatric illness, with employment in the health services or allied professions, with patients seeking compensation, and with physical findings of motor paralysis, nonanatomical sensory loss, down-going plantar response, as well as normal tone and reflexes. Sometimes patients who were thought to have traumatic spinal paraplegia and who made so-called complete recoveries and walked from the spinal injury center actually suffered from hysterical paraplegia. Several cases will be presented which suggest that nonorganic causes may be present.

CASE STUDIES
Case One

A 35-year-old black man suffered pain from the T10 level inferiorly to the low back and into both buttocks. The patient also complained of some right, lower extremity numbness extending down the posterior aspect of the right thigh, right leg, and into the plantar aspect of the foot as well as the fourth and fifth toes of the right foot. The patient described the pain as tingling, tender, hot, and itching into the low back. He also felt a numbness in the right upper extremity. He attributed his injury to lifting approximately 75–100 pounds of steel at work approximately 5 years earlier, the day after which he reports having felt low back pain.

On examination the vital statistics were within normal limits. The lower extremity pulses were palpated as 4 out of 4. Straight leg raise procedure, when performed on the right at 65°, produced low back and right hip pain. Straight leg raise procedure on the left produced low back and left hip pain at 70°. The double leg raise procedure at 75° produced pain across the low back area. The muscle strengths were grade 5 for the lower extremities. The patient described a relative hypesthesia about the plantar aspect of the right foot and right buttock. The deep tendon reflexes for the patellar and Achilles were + bilaterally. The patient also described tenderness to palpation bilaterally over the following musculature: tensor fascia latae, piriformis, gluteus maximus, triceps surae, and hamstrings.

Plain film x-rays of the lumbar and thoracic region were non-contributory.

A computerized tomographic (CT) scan of the lumbar spine was normal. Laboratory evaluations including complete bloodcount and urinalysis were normal. Thermographic evaluation of the low back and lower extremities also was unremarkable (Figs. 20.1–20.3). On a scale of 0 to 10, in which zero is no pain and 10 is the greatest amount of pain, the patient, on entrance to the clinic five years after his initial injury, ranked his pain at grade 9 or 10. He was inadvertently given therapy with a myomatic instrument in which the

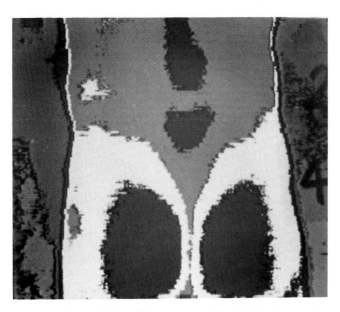

Figure 20.1. Thermogram of the low back.

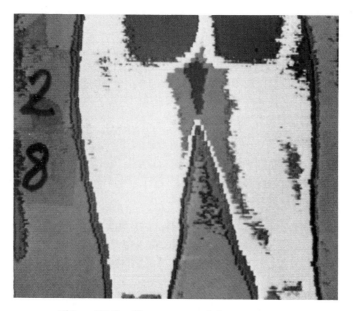

Figure 20.2. Thermogram of the upper thigh.

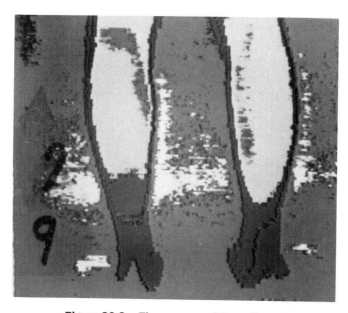

Figure 20.3. Thermogram of the calf area.

leads were not connected to render treatment. At the end of this inadvertent placebo treatment, the patient reported that his pain had diminished to grade 0–1. This placebo therapy was repeated for three visits, demonstrating consistency in the patient's response.

Case 2

A 22-year-old white woman presented to a chiropractic college center with right shoulder pain. Approximately 1 week earlier, the patient, who works for a deaf and blind school, was physically restraining a child when she was head-butted in the area of the neck and collar bone on the right side. Since the injury, the patient has noted a numbness into the right upper extremity. She had applied ice at home.

Her vital statistics were normal. Deep tendon reflexes revealed the biceps and triceps to be graded + bilaterally. All other reflexes for the upper extremities bilaterally were graded + +. Muscle strengths for the upper extremity bilateral were graded 5 +. The tricep muscle strength was graded 4 bilaterally. Standard cervical compression tests and thoracic outlet procedures were essentially unremarkable. Sensation with the cotton wisp and pinwheel revealed complete anesthesia to these modalities on the right upper extremity only. Modification of the pinwheel examination where the skin was indented substantially but was short of causing skin puncture still revealed complete anesthesia of the right upper extremity. A thermographic evaluation of the patient was symmetric. Approximately 2 weeks after the initial examination, the patient was sent to be evaluated by the Chiropractic Center's electrodiagnosis department. Median nerve, motor, sensory, and somatosensory potentials were all within normal limits.

The patient was then evaluated a second time thermographically, which still demonstrated an essentially symmetric thermogram (Figs. 20.4–20.9). She then demonstrated recovery from this disorder.

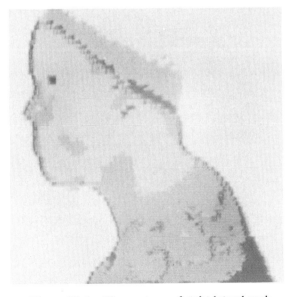

Figure 20.4. Thermogram of right lateral neck.

Figure 20.5. Thermogram of left lateral neck.

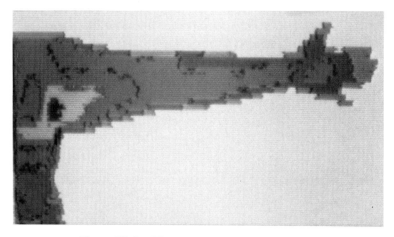

Figure 20.6. Thermogram of left anterior arm.

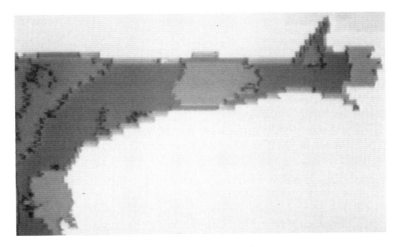

Figure 20.7. Thermogram of right posterior arm.

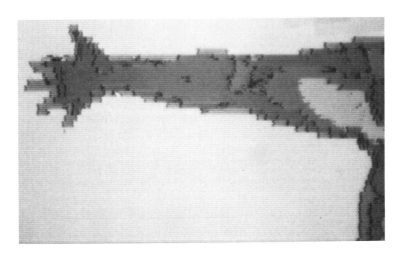

Figure 20.8. Thermogram of right anterior arm.

Figure 20.9. Thermogram of left posterior arm.

Case 3

A 33-year-old white man presented to a chiropractic college center for thermographic evaluation. He complained of numbness and aching along the posterior aspect of his neck, down to the lower lumbar region, across the anterior aspect of the chest on the left, and into the entire section of the left upper extremity. In the left lower extremity, the patient described aching and numbness from the mid-thigh anteriorly on the left side through the foot and the posterior aspect of the left thigh; he perceived numbness and pins-and-needles sensation through this entire area.

The patient described an injury which had occurred 3½ years earlier when the end of a cart was pushed into the side of his body. The pain grew worse whenever the patient lay down. Immediately following his injury, the patient was treated at an emergency room where the laceration was sutured. He then wrote a letter to his employer indicating that he could not work because he was unable to use his left arm. Five months later the employee was told his services were no longer needed. Over the 3½ years before presenting to the chiropractic center, the patient received two nerve blocks in the lower back, three nerve blocks in the neck area, three surgeries on the shoulder, electrotherapy, heat therapy, and adjustive procedures for the low back and neck. He had been given exercises and various medications to no avail.

A neurologist who evaluated the patient 10 months after the initial injury diagnosed the probability of brachial plexus involvement or left ulnar neuropathy. The patient received another examination 9 months later.

On plain film x-rays of the cervical region, the patient showed soft tissue swelling about the lateral portion of the cervical musculature on the right. Films of the right clavicle were read as essentially normal. Thoracic spine films revealed mild spondylosis of the thoracic spine in an otherwise normal study.

Thermographic evaluation revealed a generally symmetric thermograph with the following exceptions: hypothermia of the left posterior and lateral brachium and left posterior aspect of the neck. The relative hypothermicity detected in these areas was greater than a .6°C difference and is considered significant (8) (Figs. 20.10–20.12).

A physical examination revealed normal vital signs. A 2-to-3-inch scar was noted over the posterolateral aspect of the left shoulder. The pulses were normal bilaterally. Active range of motion of the cervical spine revealed 20° flexion, 0° extension, 20° left lateral flexion, 10° right lateral flexion, 45° left rotation, and 50° right rotation. Active neck flexion as well as right rotation of the neck produced low back pain. Sensation as perceived with the pinwheel, vibratory tuning fork, and cotton wisp revealed the entire right side of the body to be anesthetic and the left side to be essentially normal. Sensation with a 128 HZ tuning fork over the sternal region immediately lateral to midline on the right was completely anesthetic; a similar area on the left could also be perceived. Dynamometer testing when performed on the right revealed the following grip strengths in pounds: 110, 125, and 135. The same procedure performed on the left side revealed 0, 1, and 0 pounds. All muscle strengths on the right side of the body were grade 5 with the exception of hamstrings and peronei on the right, which were grade 4. On the left side, the

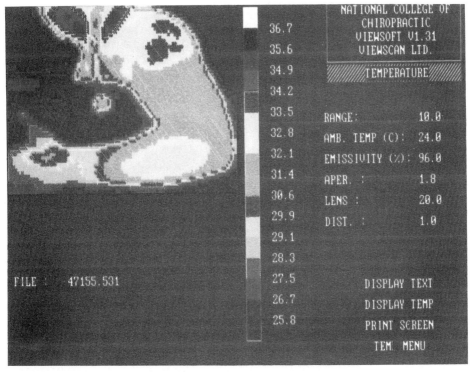

Figure 20.10. Thermogram of lateral arm.

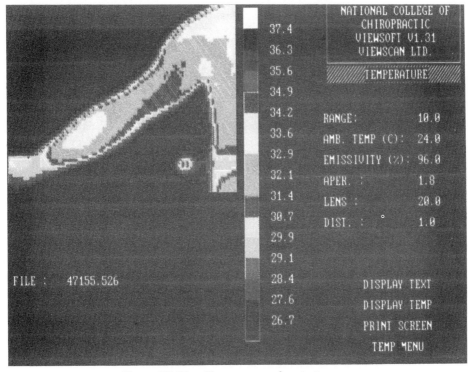

Figure 20.11. Thermogram of posterior arm.

Figure 20.12. Thermogram of posterior neck.

arm and forearm muscles were grade 4, hand muscles were grade 5, and lower extremity muscles were grade 5. Range of motion of the thoracolumbar spine revealed flexion 0°, extension 5°, right lateral flexion 15°, left lateral flexion 20°, and rotation was limited bilaterally. The straight leg raise on the right produced low back pain on the right at 70–75°. On the left the straight leg raise procedure could be performed only to a similar height with no production of back discomfort. While lying in the supine position the patient noted excruciating low back pain. No further testing was done in this position. The patient was able to perform the Bekhterev-Mendel procedure bilaterally. He demonstrated a positive Romberg's sign and was unable to perform the tandem walk procedure.

The patient scored 82 of 100 on the Oswestry examination, indicating the patient to be either bed-bound or exaggerating his symptoms.

The patient scored at least 3 penalty points on a pain pattern chart. There was a positive correlation between a normal Minnesota Multiphasic Personality Inventory and two or fewer penalty points.

DISCUSSION

In the first case of possible nonorganic problems, it would appear that the patient may have had conversion disorder or somatoform pain disorder based on the inability to substantiate a physical complaint and on reported pain reduction after inadvertent placebo treatment. His gait would change before and after treatment: he would arrive at the clinic virtually dragging his right lower extremity; he would leave following treatment walking briskly. The thermogram in this case is unremarkable.

It would appear in the second case that the patient was demonstrating a psychological disorder such as conversion. The interesting feature of this case is the total anesthesia of the right upper extremity without motor loss. There is no reason to believe she was faking the sensory deficit since testing was rigorous. Her thermogram was essentially symmetric.

The last case appears to be a combination of somatic dysfunction and somatoform pain disorder. Of interest is a comparison of this examination to that done by the two previous consultants. On first examination, the patient's grip strength was reportedly weaker on the left side. On second examination by another physician, the patient's hand power and reflexes were normal. On third examination, the patient demonstrated normal hand strengths to gross muscle testing; however, to dynamometer testing, he demonstrated gross weakness on the left. The reflexes were graded normal bilaterally. The loss of vibratory sensation when testing the sternal region only on the left is felt to be a sign of a non-organic problem (9). The thermogram is significant for the left proximal upper extremity area but does not correlate with other areas of the patient's complaint. Interestingly, he was completely unable to flex forward at the waist because of pain, yet he could assume an L position seated, simulating a 90° angle between the thoracolumbar spine and the thighs.

Thermography could be a helpful tool in helping to separate somatic dysfunction from psychogenic or malingering disorders. Controlled research with blind thermograms in random order with the two groups of patients (psychological and physical disorders) would be useful.

CONCLUSION

Nonorganic complaints often are difficult to detect accurately. In the three cases of possible nonorganic disorders presented here, the thermogram demonstrated an unremarkable study over the areas of the patients' complaints. In the final study, positive correlation was noted in a portion of the patient's area of complaint. Thermography could be a helpful tool in separating somatic dysfunction from psychogenic or malingering disorders.

References

1. Tager R: Clinical brief: Simulated disability. *J Occup Med* 27:915–916, 1985.
2. Leavitt F: Detection of simulation among persons instructed to exaggerate symptoms of low back pain. *J Occup Med* 29:229–233, 1987.
3. Leavitt F, Sweet J: Characteristics and frequency of malingering among patients with low back pain. *Pain* 25:357–364, 1986.
4. Yelin E: The myth of malingering: why individuals withdraw from work in the presence of illness. *Milbank Q* 64:622–649, 1986.
5. Pankratz L, Binder L, Wilcox L: Evaluation of an exaggerated somatosensory deficit with symptom validity testing [Letter to the editor]. *Arch Neurol* 44:798, 1987.
6. Gould R, Miller B, Goldberg M, Benson D: The validity of hysterical signs and symptoms. *J Nerv Ment Dis* 174:593–597, 1986.

7. Miller B, Benson D, Goldberg M, Gould R: Misdiagnosis of hysteria. *Am Fam Physician* 34:157–160, 1986.
8. Feldman F, Nickologg E: Normal thermographic standards for the cervical spine and upper extremities. *Skeletal Radiol* 12:235–249, 1984.
9. Dejong R, Magee K: *The Neurologic Examination*, ed 4. Hagerstown, MD., Harper and Row, 1979.

21

Meralgia Paresthetica

William Dudley, DC, DABCT

ANATOMY

The iliohypogastric nerve emits at the first lumbar (L1) as part of the lumbar plexus and produces the lateral femoral cutaneous (LFC) nerve. The distribution is the lateral thigh; the LFC is superficial and sensory (1).

CASE STUDY

A 39-year-old white salesman presented with pain at the right lateral thigh and lower back. Trauma was denied. The pain, which had existed for one week, increased while driving for brief distances; long journeys were avoided. Bed rest did not relieve the pain. In fact, the pain interrupted sleep and forced the patient to become ambulatory after two hours before being able to return to sleep. His back pain was located at the left sacroiliac area. Dry and moist heat applied to the back brought him no relief. He began limping on the right leg to favor his low back and right leg pain. Although he was able to walk, it, too, brought pain. He felt he could no longer stand erect. He admitted a prior wrist fracture but the history was unremarkable.

The patient weighed 169 pounds, his height was 74 inches, blood pressure 104/60, pulse 74, respiration 18, and his abdominal, knee, ankle and plantar reflexes were normal. There was no lower limb clonus. Both legs flexed equally. Orthopaedic tests were within normal range. Pinwheel to the lateral thigh produced some reduction of sensation on the right side. The pelvis was not level, and the tilt was to the left. His stride was restricted, and he took small steps to avoid pain. He arose from the seated position slowly and was unable to stand straight. Muscle testing was negative for any significant losses.

The x-rays were negative for fracture or osseous pathology. There was a slight left tilt to the pelvis. Comparison with prior views taken 8 years earlier showed no gross osseous changes.

One thermographic scan was performed on electronic equipment following the protocol of the International Thermographic Society (2). The thermal pattern of the lower back was consistent with spasm in that there was a general increase in the lumbar musculature emission, but there were no isolated thermal variances beyond .15°C. No dermatomal or vascular pattern could be noted in the low back or limbs, and no thermal variance beyond .2°C was seen in the limbs when compared to the opposite side. The right lateral thigh was .45°C more than the left in the area of the LFC as measured by isotherm. A review scan 3 weeks later showed no thermal variants in the back or limbs. The lateral thighs were within .2°C (Figs. 21.1 and 21.2).

DISCUSSION

Meralgia paresthetica (MP) occurs infrequently (3). The commonly described patient is one with a pendulous abdomen, yet MP often occurs in slender persons (4). Ages vary from 15 to 60 years and, again, the body conformation is often not obese. Although use of the pinwheel to the LFC can be used to diagnose MP, no quantification exists. Neither does the subjective nature of the pinwheel examination allow for good recovery substantiation.

Thermography can be used to quantify the loss or excitation of the sensory nerve (5) and to substantiate recovery. The spinal adjustment to cause restoration of MP had been to the first lumbar spine. Though not routine at follow-up, there frequently is some thermal residue loss noted (6).

The thermographic scan may reveal the existence of heated areas of the LFC; in my clinical experience, these cases are of recent occurence. More commonly, the thermogram shows the area of the LFC to have a thermal loss and is assumed to be a chronic state. The LFC is a sensory nerve and no

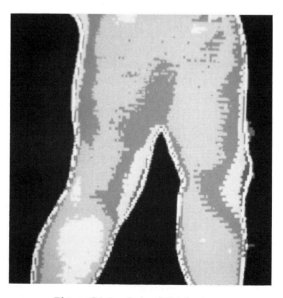

Figure 21.1. Lateral thigh view.

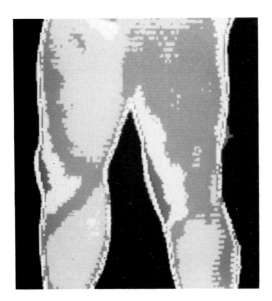

Figure 21.2. Lateral thigh view.

description is noted for motor activity. Yet in acute conditions, the patient may present with low back spasm and antalgia. It seems safe to assume that another spinal nerve branch with motor activity is also unvalued and that the area must be too deep for detection by thermography. Additional basic research may provide further insights.

It is common to discover the loss of the LFC in a routine scan, but the patient reports little or no back or leg pain. The condition often described by a patient is that some back pain had existed but is not currently symptomatic; the presenting symptom may be knee pain on the side of LFC loss. Then the thermographic scan notes loss over the LFC supplied area. Spinal correction almost always causes these cases to normalize, as seen by thermographic follow-up, even though meralgia parasthetica is noted as an entrapment. Basic research is necessary to resolve this question as well.

References

1. Chusid, JG: *Correlative Neuroanatomy and Functional Neurology*, E. Norwalk, CT, Lange Medical Publications, 1985.
2. Policy Statement, International Thermographic Society; Atlanta, June 1988.
3. Kadel RE, Godbey WD: Meralgia paresthetica: a study of incidence. *J Manipulative Physiol Ther* 6:2, 1983.
4. Gateless D, Nefcy PM, Gilroy J: Entrapment neuropathies and liquid crystal thermography. *Neurology* 36:7, 1985.
5. Gateless D, Di Gullis P, Ingall FRF, Mahmud MZ, Gilroy J: Thermography in meralgia paresthetica. *Neurology* 33:5, 1983.
6. Dudley DN: Thermography: tracking nerve traps. *ACA J Chiroprac* 21(3):63–66, 1987.

Index